PATIENT SAFETY

An Engineering Approach

PATIENT SAFETY

An Engineering Approach

B.S. DHILLON

CRC Press is an imprint of the
Taylor & Francis Group, an **informa** business

CRC Press
Taylor & Francis Group
6000 Broken Sound Parkway NW, Suite 300
Boca Raton, FL 33487-2742

First issued in paperback 2017

ISBN 13: 978-1-138-11500-2 (pbk)
ISBN 13: 978-1-4398-7386-1 (hbk)

Library of Congress Cataloging-in-Publication Data

Dhillon, B. S. (Balbir S.), 1947-
Patient safety : an engineering approach / B.S. Dhillon.
p. ; cm.
Includes bibliographical references and index.
Summary: "Each year a vast sum of money is spent on health care around the globe and patient safety has become a serious global public health issue. Filled with up-to-date information, this book demonstrates how to handle patient safety-related problems by using methods developed in the area of engineering. It contains a chapter on mathematical concepts and another chapter on introductory material on safety and human factors considered essential to understand materials presented in subsequent chapters. The author's presentation covers topics in such a manner that readers will require no previous knowledge to understand the concepts"--Provided by publisher.
ISBN 978-1-4398-7386-1 (hardcover : alk. paper)
I. Title.
[DNLM: 1. Medical Errors--prevention & control. 2. Safety Management--methods. 3. Equipment Safety--methods. 4. Models, Theoretical. 5. Patient Care--methods. 6. Quality Assurance, Health Care--methods. WB 100]

610.28'9--dc23 2011039836

Visit the Taylor & Francis Web site at
http://www.taylorandfrancis.com

and the CRC Press Web site at
http://www.crcpress.com

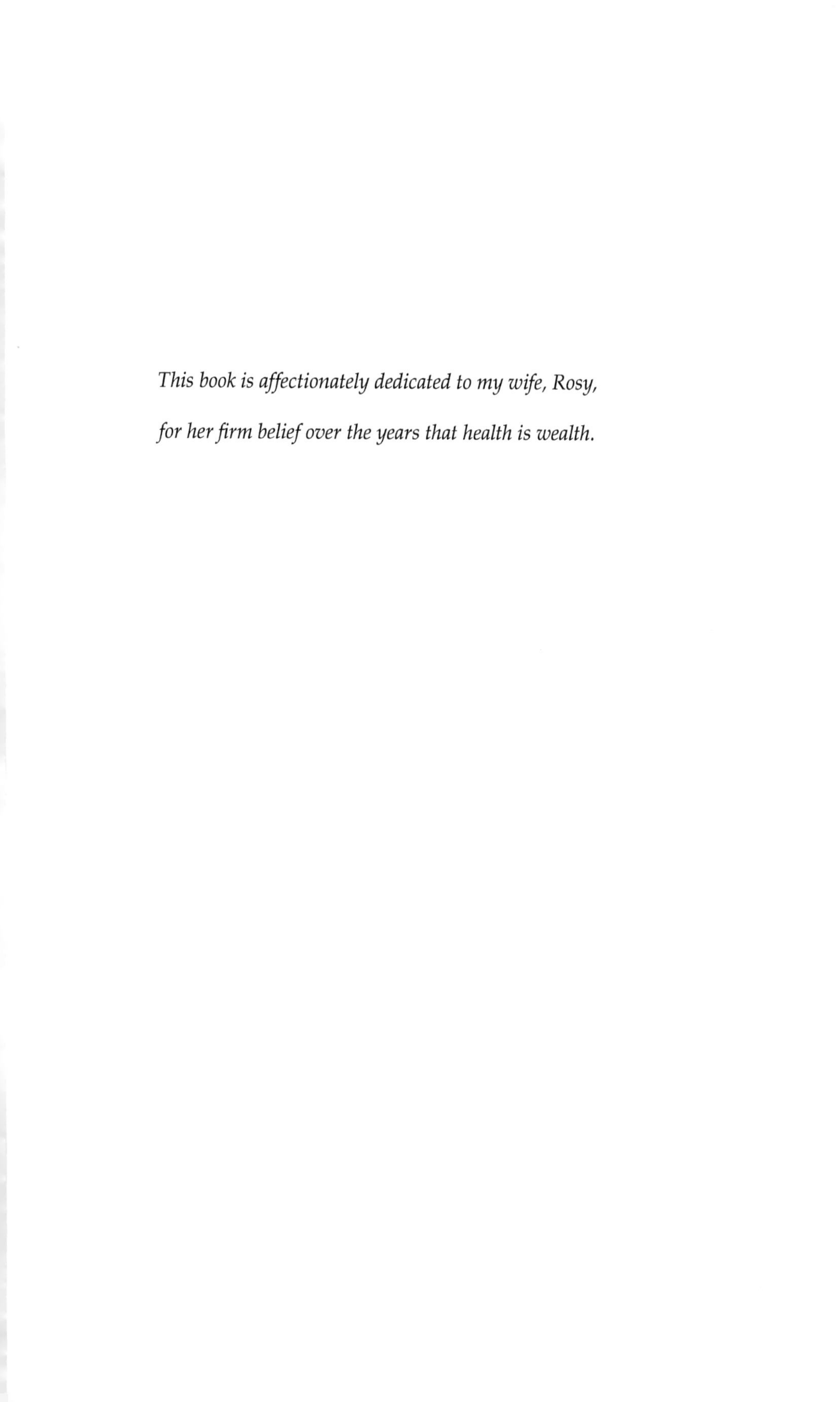

This book is affectionately dedicated to my wife, Rosy,

for her firm belief over the years that health is wealth.

Contents

Preface

Each year a vast sum of money is spent on health care around the globe, and patient safety has become a serious global public health issue because it results in millions of deaths costing billions of dollars to the world economy each year. As per the World Health Organization, in developed countries alone as many as 1 in 10 patients is harmed while receiving hospital care due to a range of errors or adverse events.

Over the years a large number of journal and conference proceedings articles on patient safety have appeared, but there are only a small number of books on the topic. In fact, to the best of the author's knowledge, there is no book that reflects a comprehensive review of published journal and conference proceedings articles on the topic and considers methods and techniques developed in the area of engineering to handle safety and human error-related problems. This causes a great deal of difficulty to information seekers on the subject, because they have to consult many different and diverse sources.

Thus, the main objective of this book is to eliminate the need to consult many different sources to obtain desired information, to provide up-to-date information on the subject, and to demonstrate to the reader how to handle patient safety-related problems by using the methods developed in the area of engineering. The book provides the source of most of the material presented, in references at the end of each chapter. This will be useful to readers who desire to delve deeper into a particular area. The book contains a chapter on mathematical concepts and another chapter on introductory material on safety and human factors considered essential to understand materials presented in subsequent chapters.

The topics covered in the volume are treated in such a manner that the reader requires no previous knowledge to understand the contents. At appropriate places the book contains examples along with their solutions, and at the end of each chapter there are numerous problems to test reader comprehension. A comprehensive list of references covering the period 1967–2011 on patient safety is provided in the appendix, to give readers a view of the intensity of developments in the area.

The book is composed of 11 chapters. Chapter 1 presents the various introductory aspects of patient safety including patient safety-related facts and figures, terms and definitions, and sources for obtaining useful information on patient safety. Chapter 2 reviews mathematical concepts considered useful to understand subsequent chapters and covers topics such as mode, median, mean deviation, Boolean algebra laws, probability definition and properties, Laplace transforms, and probability distributions.

Chapter 3 presents introductory aspects of safety and human factors. Chapter 4 is devoted to methods considered useful to perform patient safety analysis. These methods include failure modes and effect analysis (FMEA), fault tree analysis (FTA), root cause analysis (RCA), hazard and operability analysis (HAZOP), six sigma methodology, preliminary hazard analysis (PHA), interface safety analysis (ISA), and job safety analysis (JSA). Patient safety basics are presented in Chapter 5. This chapter covers such topics as patient safety goals, causes of patient injuries, patient safety culture, factors contributing to patient safety culture, safe practices for better health care, and patient safety indicators and their selection.

Chapter 6 is devoted to medication and drug safety and errors. Some of the topics covered in the chapter are drug safety in emergency departments, medication safety in operating rooms, prescribing faults and prescription errors, medication-use safety indicators, medication error types, and guidelines for reducing the occurrence of medication errors. Chapter 7 presents various important aspects of health workers' role in safety and falls including relationships between nursing workload and patient safety, health care workers' hazards, inpatient bed falls, wheelchair-related falls, and fall prevention recommendations for acute and long-term care.

Chapter 8 is devoted to human error in various medical areas and other related information. Some of the topics covered in the chapter are human error in anesthesia, emergency medicine, operating rooms, and intensive care units; factors contributing to human error in surgical pathology; guidelines for preventing the occurrence of medical errors; and health care human error reporting systems. Chapter 9 covers various important aspects of medical device safety and errors including types of medical device safety, medical device hardware and software safety, essential safety requirements for medical devices, medical devices with a high incidence of human error, medical device operator errors, and general approach to human factors during the medical device development process for reducing human errors.

Chapter 10 is devoted to medical device usability. It covers topics such as medical device users and use environments, medical device user interfaces, an approach to develop medical devices' effective user interfaces, guidelines to reduce medical device user interface-related errors, guidelines for designing hand-operated devices with respect to cumulative trauma disorder, and useful documents for improving usability of medical devices. Chapter 11 presents three important topics relating to patient safety: patient safety organizations, data sources, and mathematical models for performing probabilistic patient safety analysis.

The book is valuable for health care professionals, administrators, and students; biomedical engineers and biomedical engineering graduate students; health care researchers and instructors; safety and human factors professionals; and engineers-at-large and associated professionals concerned with medical equipment/devices.

The author is deeply indebted to many individuals, including friends, colleagues, and students for their input. The unseen contributions of my children, Jasmine and Mark, are also appreciated. Last, but not the least, I thank my wife and friend, Rosy, for typing this entire book and for her timely help in proofreading.

B. S. Dhillon
Ottawa, Ontario

Author

B. S. Dhillon, PhD is a professor of engineering management in the Department of Mechanical Engineering at the University of Ottawa. He has served as a chairman/director of the Mechanical Engineering Department/Engineering Management Program for over 10 years at the same institution. Dr. Dhillon is the founder of the probability distribution named Dhillon Distribution. He has published over 355 (i.e., 211 journal and 144 conference proceedings) articles on reliability engineering, maintainability, safety, engineering management, and so on. He is or has been on the editorial boards of 11 international scientific journals. In addition, Dr. Dhillon has written 38 books on various aspects of health care, engineering management, design, reliability, safety, and quality published by Wiley (1981), Van Nostrand (1982), Butterworth (1983), Marcel Dekker (1984), Pergamon (1986), and so on. His books are being used in over 100 countries and many are translated into languages such as German, Russian, Chinese, and Persian (Iranian).

Dr. Dhillon has served as general chairman of two international conferences on reliability and quality control held in Los Angeles and Paris in 1987. He has also served as a consultant to various organizations and bodies and has many years of experience in the industrial sector. At the University of Ottawa, he has been teaching reliability, quality, engineering management, design, and related areas for over 31 years, and he has also lectured in over 50 countries, including keynote addresses at various international scientific conferences held in North America, Europe, Asia, and Africa. In March 2004, Dr. Dhillon was a distinguished speaker at the Conference and Workshop on Surgical Errors (sponsored by the White House Health and Safety Committee and the Pentagon), held at the Capitol Hill (One Constitution Avenue, Washington, D.C.).

Dr. Dhillon attended the University of Wales, where he received a BS in electrical and electronic engineering and an MS in mechanical engineering. He received a PhD in industrial engineering from the University of Windsor.

1

Introduction

1.1 Background

Patient safety has become a major concern around the globe, because it results in millions of deaths costing billions of dollars each year to the world economy. In the United States alone, preventable adverse safety-related events cause 44,000–98,000 deaths and a very large number of injuries annually costing $17–29 billion to its economy each year [1].

It appears that the modern patient safety movement started in 1991 with the publication of the results of the Harvard Medical Practice Study in the *New England Journal of Medicine* [2–4]. In the study, the medical records of 30,000 patients hospitalized in acute care hospitals in New York State in 1984 were examined. In 1996, the American Medical Association announced the formation of the National Patient Safety Foundation [2]. In 1999, the National Academy of Sciences' Institute of Medicine released its report entitled "To Err Is Human: Building a Safer Health System" [5]. The report stated that medical errors are causing 44,000–98,000 preventable deaths annually in the United States. In 2001, the United States Congress appropriated $50 million per year for patient safety research to the Agency for Healthcare Research and Quality (AHRQ) [2].

Since 2001, many other developments on patient safety have taken place. An extensive list of publications on patient safety covering 1967–2011 is presented in the bibliography at the end of this book.

1.2 Patient Safety Facts and Figures

Some of the facts and figures concerned, directly or indirectly, with patient safety are as follows:

- A study of 37 million hospitalizations in the Medicare population in the United States revealed that there were about 1.14 million patient safety–related incidents [6].
- In a typical year preventable adverse events in the United States cause 44,000–98,000 deaths and cost $17–29 billion to its economy [5,7].

- Each year in the United Kingdom 0.3–1.4 million patients in the National Health Service hospital sectors are affected by adverse events [7,8].
- At any given point in time, over 1.4 million people around the globe suffer from infections acquired in hospitals [9].
- In 2008, the Agency for Healthcare Research and Quality (AHRQ) reported that preventable medical-related injuries in the United States are on the rise (i.e., about 1% a year) [10].
- The cost of mediation errors to the U.S. economy is estimated to be over $7 billion per year [11].
- In 2004, a Canadian study on adverse events reported that adverse events occurred in over 7% of hospital admissions and that 9000–24,000 Canadians die each year due to preventable medical errors [12].
- Each day, about 150 deaths occur in European Union (EU) countries due to infections acquired in hospitals [13].
- An Australian study reported that during the period 1988–1996, about 2.4% to 3.6% of all hospital admissions in the country were drug-related and 32% to 69% of them were preventable [14].
- In 2003, a study reported that 18 types of medical errors in the United States account for 2.4 million extra hospital days and $9.3 billion in excess charges annually [15].
- As per Ref. [2], the annual number of preventable adverse events suffered by hospitalized patients in the United States varies from 1.3 million to 15 million per year.
- A study revealed that a Hong Kong teaching hospital administered 16,000 anesthetics in one year and reported 125 related incidents [16]. A subsequent investigation of these incidents clearly indicated that human error was a factor in 80% of the cases [16].
- As per Ref. [17], the rates of disagreement between emergency physicians and radiologists in regard to the interpretation of radiographs vary from 8% to 11%.
- As per Ref. [18], the adverse drug events rate is 227 per 1000 resident-years in the nursing homes.
- A study of the occurrence of critical incidents in an intensive care unit during the period 1989–1999 reported that most of the incidents were the result of staff errors and not equipment failures [19].
- A study of 5612 surgical admissions to a hospital revealed that 36 adverse outcomes were the result of human errors [20].
- According to the Institute of Medicine (IOM), 1.5 million patients in the United States experience an adverse drug event per year [21].

- As per Refs. [9,12,22,23], about 10% of people who receive health care in the industrialized countries are expected to suffer to a certain degree because of preventable harm and adverse events.
- Medical Practice Studies (MPS) conducted in the United Kingdom, Canada, Australia, New Zealand, France, the Netherlands, and Denmark reported patient injury rates from 7.5% to 15% [2,12,24].
- As per the findings of the World Health Organization (WHO), at least half of all types of equipment used in many developing countries is partly usable or unusable altogether, thus leading to an increased degree of risk of harm to both patients and health care workers [22,25].
- A study of anesthetic-related incidents in operating rooms revealed that between 70% and 82% of the incidents were due to human error [26,27].

1.3 Terms and Definitions

Some frequently used terms and definitions taken from various sources are as follows [5,28–41].

- **Patient safety.** Freedom from accidental injury, and ensuring patient safety involves the creation of operational systems/processes that reduce the likelihood of error occurrence and increase the likelihood of error occurrence interception.
- **Adverse event.** An injury due to a medical-related intervention.
- **Safe practices.** Those practices that have proved to lower the risk of the occurrence of adverse events related to exposure to medical care across a range of conditions or diagnoses.
- **Risk.** Degree of chance, probability, or possibility of loss.
- **Health care.** Services provided to communities or individuals to monitor, restore, promote, or maintain health.
- **Quality of care.** The level to which delivered health services satisfy established professional standards and judgments of value to all types of consumers.
- **Medication error.** Any preventable event that may cause or lead to wrong medication use or patient harm while the medication is in the control of a health care professional, a consumer, or a patient.
- **Health care organization.** An entity that coordinates, provides, and/or insures medical-associated services for public/people.

- **Human factors.** A study of the interrelationships between humans, the tools they use, and the surrounding day-to-day environment in which they work and live.
- **Human error.** The failure to perform a given task (or the performance of a forbidden action) that could lead to disruption of scheduled operations or damage to property and equipment.
- **Failure.** The inability of an item to operate within the specified guidelines.
- **Fault.** An immediate cause of a failure.
- **Anesthesiology.** A branch of medicine that deals with the processes of rendering patients insensitive to various types of pain during the surgical process or when faced with chronic/acute pain states.
- **Human reliability.** The probability of performing a given task successfully by humans at any required stage in system operation within the specified minimum time limit (if the time requirement is stated).
- **Medical device.** Any instrument, apparatus, implant, machine, in vitro reagent, contrivance, implement, or other similar or related article, including any part, component, or accessory, which is intended for application in diagnosing diseases or other conditions or in the treatment, mitigation, prevention, or cure of disease or intended to affect the structure or any function of the body.
- **Operator error.** An error that occurs when an item/system/equipment operator does not follow correct or proper procedures.
- **Medical technology.** Drugs, equipment, methods, and procedures used by professionals working in health care, in delivering medical care to people and the systems within which such care is delivered.
- **Mission time.** The time during which the item is carrying out its specified mission.
- **Hazardous situation.** A condition with a potential to threaten human health, life, properties, or the environment.
- **Quality.** The extent to which the properties of a product or service generate/produce a specified outcome.
- **Mean time to failure.** In the case of exponentially distributed times to failure, the sum of the operating time of specified items over the total number of failures.
- **Human performance.** A measure of actions and failures under given conditions.
- **Continuous task.** A task that involves some kind of tracking activity (e.g., monitoring a changing condition or situation).

1.4 Useful Information on Patient Safety

Some sources for obtaining, directly or indirectly, patient safety-related information are presented below, categorized by type.

1.4.1 Journals

- *Journal of Patient Safety*
- *Journal of Drug, Healthcare and Patient Safety*
- *Patient Safety in Surgery*
- *Joint Commission Journal on Quality and Patient Safety*
- *Patient Safety and Quality Healthcare*
- *Drug Safety*
- *Quality and Safety in Healthcare*
- *American Journal of Medical Quality*
- *International Journal of Healthcare Quality Assurance*
- *Journal of Quality Clinical Practice*
- *Journal of the American Medical Association*
- *New England Journal of Medicine*
- *The Lancet*
- *British Medical Journal*
- *Journal of Professional Nursing*
- *Canadian Medical Journal*
- *British Journal of Anesthesia*
- *Medical Device and Diagnostic Industry*
- *Journal of General Internal Medicine*
- *European Journal of Anesthesiology*
- *South African Medical Journal*
- *Journal of the American College of Surgeons*
- *Journal of Royal Society of Medicine*
- *Rhode Island Medical Journal*
- *American Family Physician*

1.4.2 Conference Proceedings

- Proceedings of the Annual National Patient Safety Foundation Congress

- Proceedings of the Annual National Forum on Quality Improvement in Health Care
- Proceedings of the Annual Nursing Leadership Congress: "Building the Foundation for a Culture of Safety"
- Proceedings of the International Forum on Quality and Safety in Health Care, 2007.
- Proceedings of the Annual Maryland Patient Safety Conference
- Proceedings of the Patient Safety Congress, UK, 2008–2011.
- Proceedings of the Second Annenberg Conference on Enhancing Patient Safety and Reducing Errors in Healthcare, 1998.
- Proceedings of the First Workshop on Human Error and Clinical Systems (HEC'99), 1999.
- Proceedings of the First Symposium on Human Factors in Medical Devices, 1999.
- Proceedings of the Annual Human Factors Society Conferences

1.4.3 Books

- Spath, P. L., Editor, *Error Reduction in Health Care,* Jossey-Bass, San Francisco, California, 1999.
- Dhillon, B. S., *Human Reliability and Error in Medical System,* World Scientific Publishing, River Edge, New Jersey, 2003.
- Dhillon, B. S., *Reliability Technology, Human Error, and Quality in Health Care,* CRC Press, Boca Raton, Florida, 2008.
- Dhillon, B. S., *Medical Device Reliability and Associated Areas,* CRC Press, Boca Raton, Florida, 2000.
- Youngberg, B. J., Hatlie, M., Editors, *The Patient Handbook,* Jones and Bartlett, Sudbury, Massachusetts, 2004.
- Newhouse, R. P., Poe, S., Editors, *Measuring Patient Safety,* Jones and Bartlett, Sudbury, Massachusetts, 2005.
- Vance, J. E., *A Guide to Patient Safety in the Medical Practice,* American Medical Association, Chicago, 2008.
- Croskerry, et al., Editors, *Patient Safety in Emergency Medicine,* Wolters Kluwer Health/Lippincott Williams and Wilkins, Philadelphia, 2009.
- Sullivan, J. M., Martin, R. H., *Patient Safety Handbook,* American Bar Association, Chicago, 2008.
- Currie, L., Editor, *Understanding Patient Safety,* Quay Books, London, 2007.
- Bogner, M. S., Editor, *Human Error in Medicine,* Lawrence Erlbaum Associates, Hillsdale, New Jersey, 1994.

- Kohn, L. T., Corrigan, J. M., Donaldson, M. S., Editor, *To Err Is Human: Building a Safer Health System,* National Academy Press, Washington, D.C., 1999.
- Stamatis, D. H., *Total Quality Management in Healthcare,* Irwin Professional Publishing, Chicago, 1996.
- Youngson, R. M, Schott, I., *Medical Blunders: Amazing True Stories of Mad, Bad, and Dangerous Doctors,* New York University Press, New York, 1996.
- Mulcahy, L., Lloyd-Bostock, S. M., Rosenthal, M. M., *Medical Mishaps: Pieces of the Puzzle,* Taylor and Francis, New York, 1999.
- Rosenthal, M. M., Sutcliffe, K. M., Editors, *Medical Error: What Do We Know? What Do We Do?,* John Wiley and Sons, New York, 2002.
- Graham, N. O., Editor, *Quality in Health Care,* Aspen Publishers, Gaithersburg, Maryland, 1995.
- McLaughlin, C. P., Kaluzny, A. D., Editor, *Continuous Quality Improvement in Health Care,* Jones and Bartlett, Boston, 2006.

1.4.4 Organizations

- World Alliance for Patient Safety, c/o World Health Organization, 20 Avenue Appia CH-1211, Geneva, Switzerland.
- National Patient Safety Foundation, 132 Mass Moca Way, North Adams, Massachusetts, 01247.
- Emergency Care Research Institute (ECRI), 5200 Butler Parkway, Plymouth Meeting, Pennsylvania 19462.
- National Institute for Occupational Safety and Health, 395 E Street SW, Washington, D.C.
- National Quality Forum, 601 13th Street NW, Suite 500 North, Washington, D.C.
- American Medical Association, 515 N. State Street, Chicago, Illinois 60610.
- Institute for Safe Medication Practices, 1800 Byberry Road, Suite 810, Huntingdon Valley, Pennsylvania 19006.
- National Patient Safety Agency, 4-8 Maple Street, London W1T 5HD, United Kingdom.
- American Hospital Association, 1 North Franklin, Chicago, Illinois 60606; 325 7th Street NW, Washington, D.C.
- Agency for Healthcare Research and Quality, 500 Gaither Road, Rockville, Maryland 20357-0001.
- Institute of Medicine, 2001 Wisconsin Avenue NW, Washington, D.C. 20418.

- Australian Patient Safety Foundation, P.O. Box 400, Adelaide 5001, Australia.
- Safety and Quality in Health Care, Health Care Division, Department of Health, P.O. Box 8172, Stirling Street, Perth, Australia.
- Canadian Patient Safety Institute, Suite 1414, 10235-01 Street, Edmonton, Alberta, Canada.
- Australian Medical Association, 42 Macquaire Street, Barton, ACT (Australian Capital Territory) 2600, Australia.
- The Joint Commission on Accreditation of Healthcare Organizations (JCAHO), 1 Renaissance Blvd., Oakbrook Terrace, Illinois.
- Canadian Medical Association, 1867 Alta Vista Drive, Ottawa, Ontario KlG 3Y6, Canada.
- U.S. Food and Drug Administration (FDA), 5600 Fishers Lane, Rockville, Maryland 20857-0001.
- British Medical Association (BMA), BMA House, Tavistock Square, London WC1 H9JP, United Kingdom.

1.5 Scope of the Book

Patient safety has become a major concern throughout the world, because it results in millions of deaths costing billions of dollars each year to the world economy.

Over the years a large number of journal and conference proceedings articles on patient safety have appeared, but there are only a small number of books on the topic. In fact, to the best of the author's knowledge, there is no book that reflects a comprehensive review of published journal and conference proceedings articles on patient safety and considers methods and techniques developed in the area of engineering to handle safety and human error-related problems. Thus, the main objective of this book is to eliminate the need to consult many different sources to obtain desired information, to provide up-to-date information on the subject, and to demonstrate how to handle patient safety-related problems by using the methods developed in the area of engineering. This book is an attempt to satisfy these specific needs.

Previous knowledge is not generally required to understand the material covered in this book, because the book contains a chapter on mathematical concepts and another chapter on introductory material on safety and human factors, considered essential to understand material presented in subsequent chapters. The book will be useful to many individuals including health care professionals, administrators, and students; biomedical engineering graduate students; health care researchers and instructors; safety and human factors professionals; and engineers-at-large and associated professionals concerned with medical equipment/device.

1.6 Problems

1. Write an essay on patient safety.
2. List the five most important facts and figures concerned with patient safety.
3. What is the difference between human error and adverse event?
4. What is a medical device?
5. Define the following five terms:
    a. Patient safety
    b. Medication error
    c. Safe practices
    d. Operator error
    e. Health care organization
6. List five of the most important journals to obtain patient safety–related information.
7. List four of the most important organizations for obtaining patient safety–related information.
8. Define the following terms:
    - Quality of care
    - Anesthesiology
    - Human reliability
    - Human performance
9. Compare safe medical practices with the quality of care.
10. Define the following terms:
    - Mean time to failure
    - Hazardous situation
    - Mission time

1.7 References

1. Brall, A., Human Reliability Issues in Medical Care—A Customer Viewpoint, *Journal of American Medical Association*, Vol. 28, No. 5, 2006, pp. 46–50.

2. Leape, L. L., Scope of Problem and History of Patient Safety, *Obstetrics and Gynecology Clinics of North America*, Vol. 35, 2008, pp. 1–10.
3. Brennan, T. A., Leape, L. L., Laird, N. M., et al., Incidence of Adverse Events and Negligence in Hospitalized Patients: Results from the Harvard Medical Practice Study I, *New England Journal of Medicine*, Vol. 324, No. 6, 1991, pp. 370–376.
4. Leape, L. L., Brennan, T. A., Laird, N. M., et al., The Nature of Adverse Events in Hospitalized Patients: Results from the Harvard Medical Practice Study II, *New England Journal of Medicine*, Vol. 324, No. 6, 1991, pp. 377–384.
5. Kohn, K. T., Corrigan, J. M., Donaldson, M. S., Editors, *To Err Is Human: Building a Safer Health System*, National Academy Press, Washington, D.C., 1999.
6. *Patient Safety in American Hospitals*, Report, Health Grades, Inc., Golden, Colorado, July 2004.
7. Fukuda, H., Imanaka, Y., Hayashida, K., Cost of Hospital-Wide Activities to Improve Patient Safety and Infection Control: A Multi-Centre Study in Japan, *Health Policy*, Vol. 87, 2008, pp. 100–111.
8. Warburton, R. N., Patient Safety: How Much Is Enough?, *Health Policy*, Vol. 71, 2005, pp. 223–232.
9. *Global Patient Safety Challenge: 2005–2006*, Report, World Health Organization, Geneva, Switzerland.
10. Sorra, J., Famolaro, T., et al., *Hospital Survey on Patient Safety Culture 2008 Comparative Database Report*, AHRQ Publication No. 08-0039, Agency for Healthcare Research and Quality (AHRQ), Rockville, Maryland, 2008.
11. Wechsler, J., Manufacturers Challenged to Reduce Medication Errors, *Pharmaceutical Technology*, February 2000, pp. 14–22.
12. Baker, G. R., Norton, P. G., et al., The Canadian Adverse Events Study: The Incidence of Adverse Events among Hospital Patients in Canada, *Canadian Medical Association Journal*, Vol. 170, No. 11, 2004, pp. 1678–1685.
13. *Patient Safety*, Health First Europe, Chausse de Wavre 214 D, 1050 Brussels, Belgium, 2007.
14. Roughead, E. E., Gilbert, A. L., Primrose, J. G., Sansom, L. N., Drug-Related Hospital Admissions: A Review of Australian Studies Published 1988–1996, *Medical Journal of Australia*, Vol. 168, 1998, pp. 405–408.
15. Zhan, C., Miller, M. R., Excess Length of Stay, Charges, Mortality Attributable to Medical Injuries during Hospitalization, *JAMA*, Vol. 290, No. 14, 2003, pp. 1868–1874.
16. Short, T. G., O'Regan, A., Lew, J., Oh, T. E., Critical Incident Reporting in a Anesthetic Department Quality Assurance Programme, *Anesthesia*, Vol. 47, 1992, pp. 3–7.
17. Espinosa, J. A., Nolan, T. W., Reducing Errors Made by Emergency Physicians in Interpreting Radiographs: Longitudinal Study, *British Medical Journal*, Vol. 320, 2000, pp. 737–740.
18. Gurwitz, J. H., Field, T., Avorn, J., et al., Incidence and Preventability of Adverse Drug Events in the Nursing Home Setting, *American Journal of Medicine*, Vol. 109, No. 2, 2000, pp. 87–94.
19. Wright, D., *Critical Incident Reporting in an Intensive Care Unit*, Report, Western General Hospital, Edinburgh, Scotland, UK, 2001.
20. Couch, N. P., et al., The High Cost of Low-Frequency Events: The Anatomy and Economics of Surgical Mishaps, *New England Journal of Medicine*, Vol. 304, 1981, pp. 634–637.

21. Aspden, P., Wolcott, J., Bootman, J. L., et al., *Preventing Medication Errors*, National Academy Press, Washington, D.C., 2007.
22. Donaldson, L., Philip, P., *Patient Safety: A Global Priority*, Bulletin of the World Health Organization (Ref. No. 04-015776-82), December 2004, pp. 892–893.
23. Vincent, C., Neale, G., Woloshynowych, A., Adverse Events in British Hospitals: Preliminary Retrospective Record Review, *British Medical Journal*, Vol. 322, 2001, pp. 517–519.
24. Wilson, R., Runciman, W., Gibberd, R., et al., The Quality in Australian Health Care Study, *Medical Journal of Australia*, Vol. 163, 1995, pp. 458–471.
25. Issakov, A., Health Care Equipment: A WHO Perspective, in *Medical Devices: International Perspective on Health and Safety*, edited by C. W. G. Von Grutting, Elsevier, Amsterdam, 1994, pp. 48–53.
26. Chopra, V., Bovill, J. G., Spierdijk, J., Koornneef, F., Reported Significant Observations during Anesthesia: Perspective Analysis over an 18-month Period, *British Journal of Anesthesia*, Vol. 68, 1992, pp. 13–17.
27. Cook, R. I., Woods, D. D., Operating at the Sharp End: The Complexity of Human Error, in *Human Error in Medicine*, edited by M. S. Bogner, Lawrence Erlbaum Associates, Hillsdale, New Jersey, 1994, pp. 225–309.
28. *Glossary of Terms Commonly Used in Healthcare*, Prepared by the Academy of Health, Suite 701-L, 1801 K St. NW, Washington, D.C., 2004.
29. Battles, J. B., Lilford, R. J., Organizing Patient Safety Research to Identify Risks and Hazards, *Quality and Safety in Health Care*, Vol. 12, 2003, pp. ii2–ii7.
30. Federal Food, Drug, and Cosmetic Act, As Amended, Section 201 (h), U.S. Government Printing Office, Washington, D.C., 1993.
31. Fries, R. C., *Reliable Design of Medical Devices*, Marcel Dekker, New York, 1997.
32. Graham, N. O., Editor, *Quality in Healthcare: Theory, Application, and Evaluation*, Aspen Publishers, Gaithersburg, Maryland, 1995.
33. *Definitions of Effectiveness, Terms for Reliability, Maintainability, Human Factors, and Safety*, MIL-STD-721, U.S. Department of Defense, Washington, D.C.
34. Dhillon, B. S., *Human Reliability and Error in Medical System*, World Scientific Publishing, River Edge, New Jersey, 2003.
35. Dhillon, B. S., *Reliability Technology, Human Error, and Quality in Health Care*, CRC Press, Boca Raton, Florida, 2008.
36. Dhillon, B. S., *Engineering Safety: Fundamentals, Techniques, and Applications*, World Scientific Publishing, River Edge, New Jersey, 2003.
37. Fries, R. C., *Medical Device Quality Assurance and Regulatory Compliance*, Marcel Dekker, New York, 1998.
38. Omdahl, T. P., Editor, *Reliability, Availability, and Maintainability (RAM) Dictionary*, ASQC Quality Press, Milwaukee, Wisconsin, 1988.
39. McKenna, T., Oliverson, R., *Glossary of Reliability and Maintenance Terms*, Gulf Publishing Company, Houston, Texas, 1997.
40. Naresky, J. J., Reliability Definitions, *IEEE Transactions on Reliability*, Vol. 19, 1970, pp. 198–200.
41. Von Alven, W. H., Editor, *Reliability Engineering*, Prentice Hall, Englewood Cliffs, New Jersey, 1964.

2

Patient Safety Mathematics

2.1 Introduction

Mathematics plays an important role in finding solutions to various types of engineering- and science-related problems. Its application ranges from finding solutions to interplanetary-related problems to designing safe and reliable equipment for use in the area of health care. The history of mathematics may be traced back to the development of our currently used number symbols over 2200 years ago. The very first evidence of the use of these symbols is found on stone columns erected in 250 B.C. by the Scythian emperor Asoka of India [1]. However, the thinking on the probability concept is relatively new and it may only be traced back to the writings of Girolamo Cardano (1501–1576) in which he considered some interesting probability-related issues [1,2].

Today, various mathematics and probability concepts are being used to study various types of safety-related problems. For example, probability distributions are used to represent times to human error in performing various types of time-continuous tasks in the area of safety [3–7]. In addition, the Markov method is used to conduct human performance reliability analysis in regard to engineering systems safety [7–9].

This chapter presents various mathematical concepts considered useful to perform mathematical analysis in the area of patient safety.

2.2 Range, Mode, Median, Arithmetic Mean, and Mean Deviation

A set of given patient safety or other data is only useful if it is analyzed properly. Certain characteristics of data play a key role in describing the nature of a given set of data, thus allowing people to make better decisions. This section presents a number of statistical measures considered useful to perform patient safety-related analysis.

2.2.1 Range

This is a good measure of variation or dispersion. It may be described as the difference between the largest and the smallest values in a given set of data.

Example 2.1

Assume that the following values represent the number of monthly patient safety-related problems in a health care facility over a period of 12 months:

4, 5, 7, 15, 3, 20, 9, 11, 17, 22, 8, and 13

Find the range of the given data values.

By examining the given data values, we conclude that the highest and the smallest values are 22 and 3, respectively. Thus, the range, R, of the given data set is

$$\begin{aligned} R &= V_h - V_l \\ &= 22 - 3 \\ &= 19 \end{aligned} \tag{2.1}$$

where

V_h = the highest data value
V_l = the lowest data value

2.2.2 Mode

This is the most frequently occurring value in a given data set. However, in a set of values the mode may not exist, and even when it does exist it may not be unique.

Example 2.2

Assume that the following values represent the number of monthly patient safety problems occurring in a health care facility over a period of 12 months:

10, 4, 3, 12, 10, 15, 5, 10, 18, 17, 6, and 19

Find the mode of the above given data values.

Thus the mode of the above given data values is 10 (10 problems occurred 3 times in 12 months).

2.2.3 Median

The median of a set of values arranged in order of magnitude (i.e., lowest to highest) is the middle value or the average of two middle values.

Example 2.3

Assume that the following values represent the number of monthly patient safety-related problems occurring in a health care facility during an 11-month period: 20, 20, 5, 30, 25, 27, 35, 15, 4, 9, and 18. Find the median of the given set of values.

By arranging the given data values in order of magnitude, we obtain

4, 5, 9, 10, 15, 18, 20, 25, 27, 30, 35

Thus the median (i.e., the middle value) of the given set of data values is 18.

2.2.4 Arithmetic Mean

This is expressed by

$$m = \frac{\sum_{i=1}^{k} x_i}{k} \tag{2.2}$$

where

m = the mean value.

k = the total number of data values.

x_i = the data value i, for $i = 1, 2, 3, \ldots, k$.

Example 2.4

Assume that the following values represent the number of monthly patient safety-related incidents in a health care facility during a 9-month period: 12, 10, 8, 6, 4, 15, 9, 7, and 20. Calculate the average or mean number of patient safety-related incidents per month.

By inserting the given data values into Equation (2.2), we obtain

$$m = \frac{12+10+8+6+4+15+9+7+20}{9}$$

$$= 10.1 \text{ patient safety incidents/month}$$

Thus the average number of patient safety-related incidents per month is 10.1. In other words, the arithmetic mean of the given data values is 10.1 patient safety-related incidents per month.

2.2.5 Mean Deviation

This is expressed by

$$\theta = \frac{\sum_{i=1}^{k} |y_i - m|}{k} \tag{2.3}$$

where

θ = the mean deviation.
m = the mean or average value of the given set of data.
k = the number of data points values in a given data set.
y_i = the ith data value, for $i = 1, 2, 3, \ldots, k$.
$|y_i - m|$ = the absolute value of the deviation of y_i from m.

Example 2.5

Calculate the mean deviation of the set of data given in Example 2.4. Thus from Example 2.4 the calculated mean value of the data set is m = 10.1 patient safety incidents/month.

By using the above value and the given data in Equation (2.3), we obtain

$$\theta = \frac{[|12-10.1| + |10-10.1| + |8-10.1| + |6-10.1| + |4-10.1| + |15-10.1| + |9-10.1| + |7-10.1| + |20-10.1|]}{9}$$
$$= 3.7$$

Thus the mean deviation of the given set of data is 3.7.

2.3 Sets and Boolean Algebra Laws

Sets and Boolean algebra laws play an important role in probabilistic safety analysis, and both are presented here [10].

2.3.1 Sets

A set may be defined as any well-defined list, class, or collection of objects. Objects are normally referred to as the elements or members of the set. Usually, sets are denoted by capital letters such as X, Y, A, and B, and their elements by the lowercase letters such as d, n, m, and t.

Two basic set operations are as follows:

- **Union of sets.** This is denoted by either the symbol $\cup$ or +. For example, if $X + Y = Z$, it means that all the elements in set X or in set Y or in both sets (i.e., X and Y) are contained in set Z.
- **Intersection of Sets.** This is denoted by a dot (•) or no dot at all or $\cap$. For example, if $A \cap B = C$, it simply means that set C contains all elements that belong to both sets A and B. If there are no common elements between A and B (i.e., $A \cap B = 0$) then these two sets are called mutually exclusive sets or events.

2.3.2 Boolean Algebra Laws

Boolean algebra is named after the mathematician George Boole (1813–1864) and plays an important role in probabilistic safety analysis. Some of its laws are as follows [10–12]:

$$\bullet\ X\,Y = Y\,X \tag{2.4}$$

where
X = a set or an event.
Y = a set or an event.

$$\bullet\ XX = X \tag{2.5}$$

$$\bullet\ X + X = X \tag{2.6}$$

$$\bullet\ X + XY = X \tag{2.7}$$

$$\bullet\ X(X + Y) = X \tag{2.8}$$

$$\bullet\ X(XY) = XY \tag{2.9}$$

$$\bullet\ X(Y + Z) = XY + XZ \tag{2.10}$$

$$\bullet\ (X + Y)(X + Z) = X + YZ \tag{2.11}$$

where
Z = a set or an event.

2.4 Probability Definition and Properties

Probability is defined as follows [13]:

$$P(X) = \lim_{n \to \infty} \left(\frac{N}{n} \right) \tag{2.12}$$

where
$P(X)$ = the probability of occurrence of event X.
N = the number of times event X occurs in n repeated experiments.

Some of the basic properties of probability are as follows [11,13]:

- The probability of occurrence of event, say, X, is

$$0 \leq P(X) \leq 1 \tag{2.13}$$

- Probability of the sample space S is

$$P(S) = 1 \tag{2.14}$$

- Probability of negation of the sample space S is

$$P(\bar{S}) = 0 \tag{2.15}$$

- The probability of occurrence and nonoccurrence of an event, say A, is

$$P(A) + P(\bar{A}) = 1 \tag{2.16}$$

where
$P(A)$ = the probability of occurrence of A.
$P(\bar{A})$ = the probability of nonoccurrence of A.

- The probability of the union of m independent events is

$$P(A_1 + A_2 + \ldots + A_m) = 1 - \prod_{i=1}^{m} (1 - P(A_i)) \tag{2.17}$$

where
$P(A_i)$ = the probability of occurrence of event A_i, for $i = 1, 2, \ldots, m$.

- The probability of the union of m mutually exclusive events is given by

$$P(A_1 + A_2 + \ldots + A_m) = \sum_{i=1}^{m} P(A_i) \tag{2.18}$$

- The probability of an intersection of m independent events is

$$P(A_1 A_2 \ldots A_m) = P(A_1)P(A_2)\ldots P(A_m) \tag{2.19}$$

Example 2.6

Assume that a medical-related task is carried out by two independent health care professionals. The task under consideration will be performed incorrectly if either of the health care professionals makes an error. Calculate the probability that the task will not be accomplished successfully, if the probability of making an error by the health care professional is 0.1.

By inserting the given data values into Equation (2.17), we get

$$\begin{aligned} P(A_1 + A_2) &= 1 - \prod_{i=1}^{2} (1 - P(A_i)) \\ &= P(A_1) + P(A_2) - P(A_1)P(A_2) \\ &= 0.1 + 0.1 - (0.1)(0.1) \\ &= 0.19 \end{aligned}$$

Thus there is a 19% chance that the medical-related task will not be accomplished successfully.

2.5 Probability Density Function, Cumulative Distribution Function, and Expected Value Definitions

These three definitions are presented below, separately [13,14].

2.5.1 Probability Density Function

For a continuous random variable, this is defined by

$$f(t) = \frac{dF(t)}{dt} \tag{2.20}$$

where

t = time (i.e., a continuous random variable).
$f(t)$ = the probability density function.
$F(t)$ = the cumulative distribution function.

Example 2.7

Assume that the human error probability at time t (i.e., cumulative distribution function) of a health care professional is expressed by

$$F(t) = 1 - e^{-\theta t} \tag{2.21}$$

where

θ = the health care professional's error rate.
$F(t)$ = the cumulative distribution function or the human error probability of the health care professional at time t.

Obtain an expression for the probability density function by using Equations (2.20) and (2.21).

By inserting Equation (2.21) into Equation (2.20), we obtain

$$\begin{aligned} f(t) &= \frac{d(1 - e^{-\theta t})}{dt} \\ &= \theta e^{-\theta t} \end{aligned} \tag{2.22}$$

Thus Equation (2.22) is the expression for the probability density function—more specifically, the probability density function representing the time to human error of the health care professional.

2.5.2 Cumulative Distribution Function

For a continuous random variable, this is expressed by

$$F(t) = \int_{-\infty}^{t} f(x)dx \tag{2.23}$$

where

t = time (i.e., a continuous random variable).
$f(x)$ = the probability density function.

For $t = \infty$, Equation (2.23) becomes

$$F(\infty) = \int_{-\infty}^{\infty} f(x)dx \\ = 1 \quad (2.24)$$

This means that the total area under the probability density curve is always equal to unity.

Example 2.8

Prove, by using Equation (2.22), the cumulative distribution function expressed by Equation (2.21).

Thus by inserting Equation (2.22) into Equation (2.23) for $t \geq 0$, we obtain

$$F(t) = \int_{0}^{t} \theta e^{-\theta x} dx \\ = 1 - e^{-\theta t} \quad (2.25)$$

Equations (2.21) and (2.25) are identical.

2.5.3 Expected Value

The expected value or mean value of a continuous random variable is defined by

$$E(t) = \mu = \int_{-\infty}^{\infty} tf(t)dt \quad (2.26)$$

where

t = a continuous random variable.
$f(t)$ = the probability density function.
$E(t)$ = the expected value of the continuous random variable t.
μ = the mean value of the continuous random variable t.

Example 2.9

For $t \geq 0$, find the mean value of the probability density function expressed by Equation (2.22).

By substituting Equation (2.22) into Equation (2.26), we obtain

$$E(t) = \mu = \int_0^{\infty} t\theta e^{-\theta t} dt = \frac{1}{\theta} \tag{2.27}$$

Thus for $t \geq 0$, the mean value of the probability density function expressed by Equation (2.22) is given by Equation (2.27). It is to be noted that in Example 2.7, θ is the health care professional's error rate. Thus Equation (2.27) is the expression for the health care professional's mean time to human error.

2.6 Probability Distributions

A large number of probability distributions have been developed in the area of mathematics to perform various types of analysis [15,16]. This section presents some of the probability distributions considered useful to perform patient safety-related analysis.

2.6.1 Exponential Distribution

This is probably the most widely used probability distribution in safety and reliability studies. The probability density function of the distribution is defined by

$$f(t) = \lambda e^{-\lambda t}, \quad \text{for } \lambda > 0,\ t \geq 0 \tag{2.28}$$

where

$f(t)$ = the probability density function.
λ = the distribution parameter. In human reliability studies, it is known as the human error rate.
t = time.

Using Equations (2.23) and (2.28), we obtain the following expression for cumulative distribution function:

$$F(t) = \int_0^{t} \lambda e^{-\lambda t} dt = 1 - e^{-\lambda t} \tag{2.29}$$

With the aid of Equations (2.26) and (2.28), the following expression for the distribution mean value was obtained:

$$\mu = E(t) = \int_0^{\infty} t\lambda e^{-\lambda t} dt$$
$$= \frac{1}{\lambda} \tag{2.30}$$

2.6.2 Weibull Distribution

This distribution can be used to represent many different physical phenomena. It was developed in the early 1950s by W. Weibull, a Swedish mechanical engineering professor [17]. The probability density function for the distribution is defined by

$$f(t) = \frac{st^{s-1}}{\beta^s} e^{-\left(\frac{t}{\beta}\right)^s}, \; s > 0, \; t \geq 0 \tag{2.31}$$

where

s and β = the distribution shape and scale parameters, respectively.
t = time.

By inserting Equation (2.31) into Equation (2.23), we obtain the following cumulative distribution function:

$$F(t) = \int_0^t \frac{st^{s-1}}{\beta^s} e^{-\left(\frac{t}{\beta}\right)^s} dt$$
$$= 1 - e^{-\left(\frac{t}{\beta}\right)^s} \tag{2.32}$$

Using Equations (2.26) and (2.31), we obtain the following equation for the distribution mean value:

$$\mu = E(t) = \int_0^{\infty} \frac{tst^{s-1}}{\beta^s} e^{-\left(\frac{t}{\beta}\right)^s} dt$$
$$= \beta\Gamma\left(1 + \frac{1}{s}\right) \tag{2.33}$$

where

Γ(·) = the gamma function and is defined by

$$\Gamma(n) = \int_0^{\infty} t^{n-1} e^{-t} dt, \quad \text{for } n > 0 \tag{2.34}$$

Finally, it is to be noted that at $s = 1$, a Weibull distribution becomes an exponential distribution.

2.6.3 Normal Distribution

This is one of the most widely known statistical distributions, sometimes called the Gaussian distribution after Carl Friedrich Gauss (1777–1855), a German mathematician. The probability density function of the distribution is defined by

$$f(t) = \frac{1}{\sigma\sqrt{2\pi}} \exp\left[-\frac{(t-\mu)^2}{2\sigma^2}\right], \quad \text{for } -\infty < t < +\infty \tag{2.35}$$

where

μ and σ = the distribution parameters (i.e., mean and standard deviation, respectively).

t = time.

Using Equations (2.23) and (2.35), we obtain the following cumulative distribution function:

$$F(t) = \frac{1}{\sigma\sqrt{2\pi}} \int_{-\infty}^{t} \exp\left[-\frac{(x-\mu)^2}{2\sigma^2}\right] dx \tag{2.36}$$

By substituting Equation (2.35) into Equation (2.26) we get

$$E(t) = \mu \tag{2.37}$$

2.7 Laplace Transform Definition, Common Laplace Transforms, and Final-Value Theorem

The Laplace transform of the function $f(t)$ is defined by

$$f(s) = \int_0^\infty f(t) e^{-st} dt \tag{2.38}$$

where

s = the Laplace transform variable.
t = the time variable.
$f(s)$ = the Laplace transform of $f(t)$.

Example 2.10

Obtain the Laplace transform of the following function:

$$f(t) = e^{-\theta t} \tag{2.39}$$

where

θ = a constant.
t = the time variable.

By substituting Equation (2.39) into Equation (2.38), we obtain

$$\begin{aligned} f(s) &= \int_0^\infty e^{-\theta t} \cdot e^{-st} dt \\ &= \int_0^\infty e^{-(\theta+s)t} dt \\ &= \left. \frac{e^{-(\theta+s)t}}{-(\theta+s)} \right|_0^\infty \\ &= \frac{1}{(\theta+s)} \end{aligned} \tag{2.40}$$

Example 2.11

Obtain the Laplace transform of the following function:

$$f(t) = 5 \tag{2.41}$$

By substituting Equation (2.41) into Equation (2.38), we obtain

TABLE 2.1

Laplace Transforms of Some Frequently Occurring Functions in Engineering Safety or Patient Safety Analysis

No.	*f(t)*	*f(s)*
1	C (a constant)	$\frac{C}{s}$
2	t^n, for $n = 0, 1, 2, \ldots$	$\frac{n!}{s^{n+1}}$
3	$e^{-\theta t}$	$\frac{1}{s+\theta}$
4	$te^{-\theta t}$	$\frac{1}{(s+\theta)^2}$
5	$\theta_1 f_1(t) + \theta_2 f_2(t)$	$\theta_1 f_1(s) + \theta_2 f_2(s)$
6	$\frac{df(t)}{dt}$	$sf(s) - f(0)$
7	$tf(t)$	$-\frac{df(s)}{ds}$

$$
\begin{aligned}
f(s) &= \int_0^\infty 5 \cdot e^{-st} dt \\
&= \left. \frac{5e^{-st}}{-s} \right|_0^\infty \\
&= \frac{5}{s}
\end{aligned}
\tag{2.42}
$$

Table 2.1 presents Laplace transforms of a number of functions considered useful to perform mathematical patient safety analysis [18,19].

The Laplace transform of the final value theorem is expressed by [18]

$$
\lim_{t \to \infty} f(t) = \lim_{s \to 0} sf(s) \tag{2.43}
$$

Example 2.12

Prove by using the following equation that the left-hand side of Equation (2.43) is equal to its right-hand side:

$$
f(t) = \frac{\mu}{(\theta+\mu)} + \frac{\theta}{(\theta+\mu)} \cdot e^{-(\theta+\mu)t} \tag{2.44}
$$

where

θ and μ are constants.

By inserting Equation (2.44) into the left-hand side of Equation (2.43), we obtain

$$\lim_{t\to\infty}\left[\frac{\mu}{(\theta+\mu)}+\frac{\theta}{(\theta+\mu)}\cdot e^{-(\theta+\mu)t}\right]=\frac{\mu}{\theta+\mu} \tag{2.45}$$

With the aid of Table 2.1, we obtain the following Laplace transforms of Equation (2.44):

$$f(s)=\frac{\mu}{(\theta+\mu)s}+\frac{\theta}{(\theta+\mu)}\cdot\frac{1}{(s+\theta+\mu)} \tag{2.46}$$

By substituting Equation (2.46) into the right-hand side of Equation (2.43), we get

$$\lim_{s\to 0}\left[\frac{\mu}{(\theta+\mu)s}+\frac{\theta}{(s+\mu)}\cdot\frac{1}{(s+\theta+\mu)}\right]=\frac{\mu}{(\theta+\mu)} \tag{2.47}$$

As the right-hand sides of Equations (2.45) and (2.47) are exactly the same, it proves that the left-hand side of Equation (2.43) is equal to its right-hand side.

2.8 Solving First-Order Differential Equations Using Laplace Transforms

In safety studies, linear first-order differential equations are sometime solved. Laplace transforms are a useful tool for solving a set of linear first-order differential equations. The application of Laplace transforms to solve a set of linear first-order differential equations describing a patient safety problem in a health care facility is demonstrated through the following example.

Example 2.13

Assume that patient safety-related incidents occurring at a health care facility are described by the following two linear first-order differential equations:

$$\frac{dP_1(t)}{dt}+\theta P_1(t)=0 \tag{2.48}$$

$$\frac{dP_2(t)}{dt}-P_1(t)\theta=0 \tag{2.49}$$

where

$P_1(t)$ = the probability that there is no patient safety-related incident at the health care facility at time t.

$P_2(t)$ = the probability that there is a patient safety-related incident at the health care facility at time t.

θ = the patient-related incident rate.

At time $t = 0$, $P_1(0) = 1$, and $P_2(0) = 0$.

Find solutions to Equations (2.48) and (2.49) by using Laplace transforms.

By taking Laplace transforms of Equations (2.48) and (2.49) and then using the given initial conditions, we obtain

$$P_1(s) = \frac{1}{(s+\theta)} \tag{2.50}$$

$$P_2(s) = \frac{\theta}{s(s+\theta)} \tag{2.51}$$

where

$P_1(s)$ = the Laplace transform of the probability that there is no patient safety-related incident at the health care facility at time t.

$P_2(s)$ = the Laplace transform of the probability that there is a patient safety-related incident at the health care facility at time t.

By taking the inverse Laplace transforms of Equations (2.50) and (2.51), we obtain

$$P_1(t) = e^{-\theta t} \tag{2.52}$$

$$P_2(t) = 1 - e^{-\theta t} \tag{2.53}$$

Thus, Equations (2.52) and (2.53) are the solutions to differential Equations (2.48) and (2.49).

Example 2.14

Assume that at a health care facility the patient safety-related incident rate is 0.002 incidents per hour. Calculate the probabilities of patient safety-related incident occurrence and nonoccurrence at the health care facility during a 200-hour time period.

By substituting the given data values into Equations (2.52) and (2.53), we get

$$P_1(200) = e^{-(0.002)(200)}$$

$$= 0.6704$$

and

$$P_2(200) = 1 - e^{-(0.002)(200)}$$
$$= 0.3296$$

Thus the probabilities of patient safety-related incident occurrence and nonoccurrence at the health care facility during a 200-hour time period are 0.3296 and 0.6704, respectively.

2.9 Problems

1. Assume that the numbers of monthly patient safety-related problems in a health care facility over the period of 9 months were 10, 15, 6, 9, 8, 7, 2, 4, and 5. Find the range of the given data values.
2. Assume that the numbers of monthly patient safety-related incidents in a health care facility over the period of 10 months were 8, 6, 7, 10, 12, 15, 4, 8, 11, and 16. Find the mode of the given data values.
3. Find the median of the given set of values in problem 1.
4. Calculate the mean of the given set of values in problem 2.
5. Calculate the mean deviation of the set of data given in problem 1.
6. Prove that the left side of Equation (2.11) is equal to its right side.
7. Assume that a medical-related task is performed by two independent health care professionals. The task under consideration will be performed incorrectly if either of the health care professionals makes an error. Calculate the probability that the task will not be accomplished successfully, if the probability of making an error by the health care professional is 0.2.
8. Define the cumulative distribution function.
9. Write down the probability density function of the exponential distribution.
10. Assume that at a health care facility the patient safety-related incident rate is 0.0007 incidents per hour. Calculate the probabilities of patient safety-related incident occurrence and nonoccurrence at the health care facility during a 150-hour time period.

2.10 References

1. Eves, H., *An Introduction to the History of Mathematics*, Holt, Rinehart, Winston, New York, 1976.
2. Owen, D. B., Editor, *History of Statistics and Probability*, Marcel Dekker, New York, 1976.
3. Askren, W. B., Regulinski, T. L., Quantifying Human Performance for Reliability Analysis of Systems, *Human Factors*, Vol. 11, 1969, pp. 393–396.
4. Regulinski, T. L., Askern, W. B., Mathematical Modeling of Human Performance Reliability, *Proceedings of the Annual Symposium on Reliability*, 1969, pp. 5–11.
5. Regulinksi, T. L., Askren, W. B., Stochastic Modeling of Human Performance Effectiveness Functions, *Proceedings of the Annual Reliability and Maintainability Symposium*, 1972, pp. 407–416.
6. Dhillon, B. S., Stochastic Models for Predicting Human Reliability, *Microelectronics and Reliability*, Vol. 25, 1982, pp. 491–496.
7. Dhillon, B. S., System Reliability Evaluation Models with Human Errors, *IEEE Transactions on Reliability*, Vol. 32, 1982, pp. 491–496.
8. Dhillon, B. S., *Engineering Safety: Fundamentals, Techniques, and Applications*, World Scientific Publishing, River Edge, New Jersey, 2003.
9. Dhillon, B. S, *Human Reliability and Error in Medical System*, World Scientific Publishing, River Edge, New Jersey, 2003.
10. Lipschutz, S., *Set Theory*, McGraw Hill Book Company, New York, 1964.
11. Lipschutz, S., *Probability*, McGraw Hill Book Company, New York, 1965.
12. *Fault Tree Handbook*, Report No. NUREG-0492, U.S. Nuclear Regulatory Commission, Washington, D.C., 1981.
13. Mann, N. R., Schafer, R. E., Singpurwalla, N. D., *Methods for Statistical Analysis of Reliability and Life Data*, John Wiley and Sons, New York, 1974.
14. Shooman, M. L., *Probabilistic Reliability: An Engineering Approach*, McGraw Hill, New York, 1968.
15. Patel, J. K., Kapadia, C. H., Owen, D. B., *Handbook of Statistical Distributions*, Marcel Dekker, New York, 1976.
16. Dhillon, B. S., Life Distributions, *IEEE Trans. on Reliability*, Vol. 30, 1981, pp. 457–460.
17. Weibull, W., A Statistical Distribution Function of Wide Applicability, *Journal of Applied Mechanics*, Vol. 18, 1951, pp. 293–297.
18. Spiegel, M. R., *Laplace Transforms*, McGraw Hill, New York, 1965.
19. Oberhettinger, F., Badii, L., *Tables of Laplace Transforms*, Springer-Verlag, New York, 1973.

3

Safety and Human Factors Basics

3.1 Introduction

The history of safety may be traced back to ancient times, but in the modern context its serious beginnings appear to be in 1868, when a patent was awarded for the first barrier safeguard [1]. In 1877 the Massachusetts legislature passed a law requiring appropriate safeguards on hazardous machinery [1,2]. In 1931 H. W. Heinrich published a book on industrial safety entitled *Industrial Accident Prevention* [3], and in 1970 the United States Congress passed the Occupational Safety and Health Act (OSHA) [1,2,4].

Today, safety is recognized as a specialized discipline and there are a large number of publications available on the topic in the forms of books, technical reports, journal and conference proceedings articles, and standards.

The history of human factors may be traced back to 1898, when Frederick W. Taylor conducted various studies to determine the most suitable design of shovels [5]. In 1918 the United States Department of Defense established laboratories to perform research on various aspects of human factors at the Brooks and Wright-Patterson Air Force Bases [6]. In 1924 the National Research Council (USA) initiated a study concerned with the various aspects of human factors including the effects of varying illumination, length of workday, and rest period on productivity at the Hawthorne Plant of Western Electric in the state of Illinois [7,8]. By 1945 human factors came to be recognized as a specialized discipline, and currently a vast number of publications are available on human factors.

This chapter presents various fundamental aspects of safety and human factors considered useful for studying patient safety.

3.2 Need for Safety and Safety-Related Facts and Figures

Today, safety has become an important issue because each year a large number of people die or get seriously injured due to various types of accidents. As per the National Safety Council (NSC), in 1996 there were 93,400 deaths and a large number of disabling injuries due to accidents in the United States [4]. The cost of these accidents was estimated to be about $121 billion.

Some of the other factors that play an important role in demanding the need for better safety are as follows [9]:

- Public pressure
- Increasing number of lawsuits
- Government regulations

Some of the directly or indirectly safety-related facts and figures are as follows:

- In the 1990s, the cost of accidents per American worker was estimated to be about $420 per year [1].
- Every year about 300,000 new cases of occupational-related diseases are reported in the United States [1,10].
- In 2000, there were about 97,300 unintentional injury deaths in the United States. The cost of these deaths to the U.S. economy was estimated to be about $512.4 billion [11].
- In 1997, three workers in the United States were awarded $5.8 million when they sued a computer equipment manufacturing company for musculoskeletal disorders (MSDs). The workers firmly believed that these disorders were the result of keyboard entry activities [12].
- Each year in the European Union countries about 5500 persons die due to workplace accidents [13].
- In a typical year in the United States about 35 million work hours are lost due to various types of accidents [14].
- Accidental work deaths in the United States during the period 1912–1993 have reduced by 81% (i.e., from 21 deaths per 100,000 persons to 4 deaths per 100,000 persons) [14].
- In 1969, the United States Department of Health, Education, and Welfare special committee reported that during a 10-year period, there were about 10,000 medical-device-related injuries in the United States and 731 resulted in deaths [15,16].
- A software error in a computer-controlled therapeutic radiation machine known as Therac 25 resulted in the deaths of two patients and a severe injury to a patient [17,18].
- In a typical year, the work-related accidental deaths by cause in the United States are poison (gas, vapor), 1.4%; water transport-related, 1.65%; poison (solid, liquid), 2.7%; air transport-related, 3%; fire-related, 3.1%; drowning, 3.2%; electric current, 3.7%; falls, 12.5%; motor vehicle-related, 37.2%; and others, 31.6% [1,14].

3.3 Common Causes of Work Injuries and Classifications of Product Hazards

There are many causes of work injuries. Some of the common ones are shown in Figure 3.1 [1]. A study performed by the NSC reported that about 31% of all work-related injuries are due to overexertion [1,2].

Professionals working in the area of safety have identified many product-related hazards. These hazards may be categorized under the following six classifications [19]:

- **Misuse- and abuse-related hazards.** These hazards are concerned with the product usage by people and past experiences indicate that the misuse of products can lead to serious injuries. Product abuse can result in hazardous situations or injuries; abuse causes include lack of proper maintenance and poor operating practices.
- **Electrical hazards.** These hazards have two main components: shock hazard and electrocution hazard. The major electrical hazard to property/product stems from electrical faults (i.e., short circuits).
- **Energy hazards.** These hazards may be categorized under two groups: kinetic energy and potential energy. The kinetic energy-related hazards pertain to parts/items that have energy because of their motion. Three examples of such parts/units are fan blades,

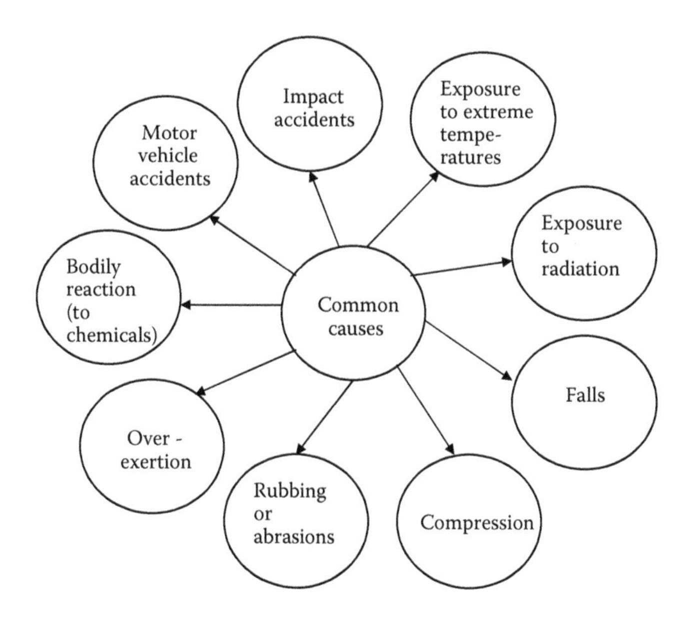

FIGURE 3.1
Common causes of work injuries.

flywheels, and loom shuttles. Any object that interferes with the motion of these parts/items can experience substantial damage. The potential energy-related hazards pertain to parts/items that store energy. Some examples of such parts/items are springs, counterbalancing weights, and electronic capacitors. During the equipment servicing process the potential energy-related hazards are very important because stored energy can lead to serious injuries when released suddenly.

- **Kinematic hazards.** These hazards are concerned with situations in which parts/items come together while moving and lead to possible cutting, pinching, or crushing of any object caught between them.
- **Environmental hazards.** These hazards may be grouped under the categories *external* and *internal*. External hazards are posed by the product/item under consideration during its life span and include disposal hazards, service-life operation hazards, and maintenance hazards. Internal hazards are concerned with the changes in the surrounding environment that lead to internally damaged product/item. This type of hazard can be eliminated or minimized by considering factors such as electromagnetic radiation, vibrations, ambient noise levels, extremes of temperatures, illumination level, and atmospheric contaminants during the design phase.
- **Human factor hazards.** These hazards are associated with poor design with respect to people—specifically, to their physical strength, computational ability, intelligence, visual acuity, length of reach, height, visual angle, weight, education, and so on.

3.4 Common Mechanical Injuries

In the industrial sector, humans interact with various types of equipment to perform tasks including drilling, stamping, punching, chipping, shaping, cutting, and stitching. During the performance of tasks such as these, some of the common types of injuries that can result are as follows [1,9]:

- **Crushing injuries:** These injuries occur when a body part is caught between two hard surfaces moving progressively together and crushing any item or object that comes between them.
- **Puncturing injuries.** These injuries occur when an object penetrates straight into the body of a person and pulls straight out. Usually, in the industrial sector, these types of injuries are associated with punching machines because they have sharp tools.

- **Shearing injuries.** These injuries are concerned with the shearing processes. In the area of manufacturing, power-driven shears are often used to perform tasks such as severing metal, plastic, paper, and elastomers. In the past, in using such equipment various tragedies such as amputation of hands/fingers have occurred.
- **Breaking injuries.** These injuries are usually associated with equipment used to deform engineering materials. Usually a break in a bone is called a fracture, and fracture is classified into categories such as oblique, simple, transverse, compound, comminuted, and complete.
- **Cutting and tearing injuries.** These injuries occur when a body part of a person comes in contact with a sharp edge. The severity of a cut or a tear depends on the degree of damage to body parts such as veins, muscles, skin, and arteries.
- **Straining and spraining injuries.** These injuries (e.g., straining of muscles or spraining of ligaments), in the industrial sector, are usually associated with the use of machines or other similar tasks.

3.5 Accident Causation Theories

Accident causation theories include the human factors theory, the domino theory, the systems theory, the combination theory, the epidemiological theory, and the accident/incident theory [1,2]. The first two of these theories are described below.

3.5.1 The Human Factors Accident Causation Theory

This theory is based on the assumption that accidents occur due to a chain of events caused by human error. It consists of the following major factors that lead to human error [1,2,20]:

- **Overload.** This is concerned with an imbalance between the capacity of a person at any time and the load he/she is carrying in a specified state. The capacity of a person is the product of factors such as stress, fatigue, natural ability, state of mind, degree of training, and physical condition. The load carried by a person is composed of tasks for which he/she has responsibility along with additional burdens resulting from environmental factors (e.g., distractions, noise), internal factors (e.g., personal problems, worry, emotional stress), and situational factors (e.g., unclear instructions, level of risk).
- **Inappropriate activities.** These activities performed by a person can be the result of human error. For example, the involved person

misjudged the degree of risk in a specified task and then carried out the task on that misjudgment.

- **Inappropriate response and incompatibility.** These two factors can also cause human errors. Three examples of inappropriate response by a person are he/she disregarded the specified procedures, he/she detected a hazardous condition but took no corrective action, and he/she removed a safeguard from a machine to increase output. Incompatibility refers to the incompatibility of a person's workstation with respect to factors such as reach, size, feel, and force. These incompatibilities can cause accidents and injuries.

3.5.2 The Domino Accident Causation Theory

This is Heinrich's accident causation theory and is operationalized in the following 10 statements known as the "Axioms of Industrial Safety" [2,3,20]:

- Most of the accidents are the result of the unsafe acts of people.
- The reasons why humans commit unsafe acts can be useful in selecting necessary corrective measures.
- Supervisors are the key people in industrial accident prevention.
- Management should assume safety responsibility aggressively because it is in the best position to achieve final results.
- The severity of an injury is largely fortuitous and the specific accident that caused it is often preventable.
- The occurrence of injuries results from a completed sequence of factors, the final one of which is the accident itself.
- An unsafe act by a person or an unsafe condition does not always immediately result in an accident/injury.
- An accident can occur only when an individual commits an unsafe act and/or there is a physical- or mechanical-related hazard.
- The most effective accident prevention approaches are analogous with the quality of productivity methods.
- There are direct and indirect costs of an accident occurrence. Some examples of the direct costs are liability claims, compensation, and medical costs.

Furthermore, as per Heinrich there are five factors in the sequence of events leading up to an accident [2,3,20]:

- **Ancestry and social environment.** In the case of this factor, it is assumed the negative character traits such as stubbornness, avariciousness, and recklessness that might lead humans to behave in an unsafe manner can be inherited through one's ancestry or acquired as a result of the social surroundings.

- **Fault of person.** In the case of this factor, it is assumed that negative traits (i.e., whether acquired or inherited) such as violent temper, recklessness, nervousness, and ignorance of safety practices constitute proximate reasons for committing unsafe acts or for the existence of mechanical or physical hazards.
- **Unsafe act/mechanical or physical hazard.** In the case of this factor it is assumed that unsafe acts such as removing safeguards and starting machinery without warning and mechanical or physical hazards such as inadequate light, absence of rail guards, and unguarded gears are the direct causes of accidents.
- **Accident.** In the case of this factor, it is assumed that events such as falls of people and striking of people by flying objects are examples of accidents that lead to injury.
- **Injury.** In the case of this factor, it is assumed that the typical injuries directly resulting from accidents include lacerations and fractures.

Finally, the two central points of the domino theory (i.e., Heinrich theory) are as follows [20]:

- The eradication of the central factor (i.e., unsafe act/hazardous condition) negates preceding factors' action and in turn prevents the occurrence of possible injuries and accidents.
- Injuries are the direct result of the action of all preceding factors (i.e., the ones described above).

3.6 Safety Management Principles

There are many principles of safety management, including the following [21,22]:

- The function of safety is to find and define the operational errors that result in accidents.
- The safety system should be made to fit organization/company culture.
- The causes that lead to unsafe behavior can be identified, controlled, and classified. Some classifications of the causes are overload, traps, and the worker's or the employee's decision to err.

- The key to successful line safety performance is management-related procedures that factor in accountability.
- There are specific sets of circumstances that can be predicted to lead to severe injuries: nonproductive activities, certain construction situations, unusual nonroutine activities, and high energy sources.
- Safety should be managed just like any other function within an organization/company. Management should direct safety by setting appropriate attainable goals, by organizing, planning, and controlling to attain them successfully.
- The three symptoms that clearly indicate something is wrong in the management system are an accident, an unsafe condition, and an unsafe act.
- Under most circumstances, unsafe behavior is normal human behavior because it is the result of normal humans or people responding to their surrounding environment. Thus, it is the management's responsibility to make appropriate changes to the environment that leads to the unsafe human behavior.
- In building a good safety system, the three main subsystems that must be considered are the managerial, the behavioral, and the physical.
- There is no single method that can effectively achieve safety in an organization. However, for a safety system to be effective, it must clearly satisfy certain criteria: be flexible, have the top-level management visibly showing its support, be perceived as positive, and involve worker participation.

3.7 Human Factors Objectives and Disciplines Contributing to Human Factors

Over the years, professionals working in the area of human factors have identified many objectives of human factors, which can be classified under the following four groups [8,23]:

- **Fundamental operational objectives.** These objectives are concerned with increasing safety, improving system performance, and reducing human errors.
- **Objectives affecting operators and users.** These objectives are concerned with improving user acceptance and ease of use; reducing physical stress, boredom, monotony, and fatigue; improving aesthetic appearance; and improving the work environment.

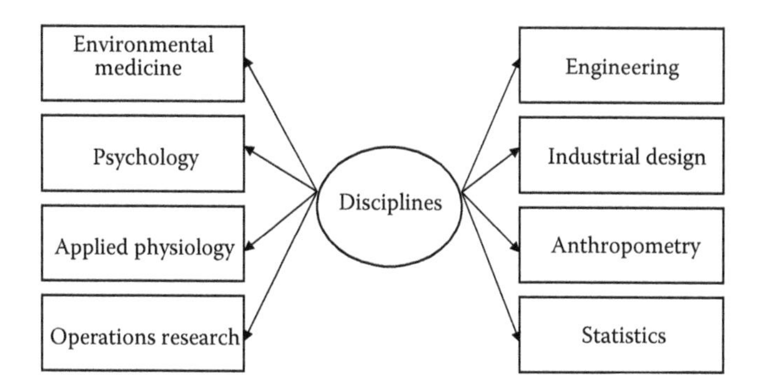

FIGURE 3.2
Disciplines contributing to human factors.

- **Objectives affecting reliability and maintainability.** These objectives are concerned with improving maintainability, reducing training needs, increasing reliability, and reducing the need for manpower.
- **Miscellaneous objectives.** These objectives are concerned with items such as reducing losses of time and equipment and improving production economy.

Many disciplines contribute to the field of human factors; some of the main ones are shown in Figure 3.2 [8,23].

3.8 Human and Machine Characteristics

Some of the main comparable human and machine characteristics are presented in Table 3.1 [8,24]. In this table, some of the characteristics may be more applicable to robots than to general machines.

3.9 General Human Behaviors

Professionals working in the area of human factors have studied humans in regard to their expected behaviors. To minimize safety-related problems, knowledge of these expected or general human behaviors is considered essential. Some of the general or typical human behaviors are as follows [25,26]:

- Most people fail to recheck specified procedures for errors.

TABLE 3.1

Comparable Human and Machine Characteristics

No.	Human Characteristic	Machine Characteristic
1	Human consistency can be low.	Machines are consistent, unless there are failures.
2	Humans require some motivation.	Machines require no motivation.
3	Humans may be absent from work due to various reasons including strikes, training, illness, and personal matters.	Machines are subject to malfunctions or failures.
4	Humans possess inductive capabilities.	Machines have poor inductive capability, but a good deductive ability.
5	Human memory could be constrained by elapsed time, but it has no capacity limitation.	The machine memory is not influenced by elapsed and absolute times.
6	Human reaction time is rather slow in comparison to that of machines.	Machines have a fast reaction time to external signals.
7	Humans are subject to fatigue that decreases with rest and increases with the number of hours worked.	Machines are free from fatigue, but require periodic maintenance.
8	Humans possess a high degree of intelligence, and are capable of applying judgments to solve unexpected problems.	Machines possess limited intelligence and judgmental capability.
9	Humans are affected by environmental factors such as temperature, hazardous materials, and noise; they also need air to breathe.	Machines are not easily affected by the environment; thus they are useful for applications in hostile environments.

- Normally, humans consider manufactured items as being safe.
- Humans are frequently reluctant to admit errors or mistakes.
- In general, humans tend to hurry.
- Humans frequently misread or overlook labels and instructions.
- Humans tend to think electrically powered switches move upward, to the right direction, and so on, to turn power on.
- Humans often respond irrationally in emergency situations.
- Humans get easily confused with unfamiliar items.
- Generally, people know quite little about their physical shortcomings.
- Humans often use their hands first to explore or test.
- People usually perform tasks while thinking about other things.
- People can get easily distracted by certain aspects of features of a product.
- Humans are generally poor estimators of speed, clearance, and distance.

- People, in the event of losing balance, tend to reach for and grab the closest item.
- A significant proportion of people become complacent after successfully handling dangerous or hazardous items/materials over a period of time.
- People tend to think that an object is small enough to get hold of and fairly light to pick up.
- People tend to think that the handles of valves rotate counterclockwise for increasing the flow of gas, liquid, or steam.
- The attention of people is drawn to items such as bright colors, flashing lights, bright lights, and loud noise.

3.10 Human Sensory Capabilities

Humans possess many useful senses: touch, hearing, sight, taste, and smell. Humans can also sense pressure, vibration, temperature, acceleration (shock), rotation, linear motion, and position. In the area of patient safety, a better understanding of sensory capabilities of humans can be useful to reduce the occurrence of human errors. Four human sensory-related capabilities are described below [8,27,28].

3.10.1 Touch

This is an important human sense because it helps to relieve the load on both ears and eyes by conveying messages to the human brain. One example of this sense is that an individual can recognize different control knobs' shapes simply by touching. The touch sensor has been used for many centuries by craft workers to detect surface irregularities and roughness in their work.

Various studies performed over the years clearly indicate that the detection accuracy improves dramatically when a person moves a thin cloth or a piece of paper over the item/object surface instead of just using bare fingers [28].

3.10.2 Sight

Sight is stimulated by the electromagnetic radiation of certain wavelengths, frequently referred to as the visible segment of the electromagnetic spectrum. The various areas of the spectrum, as seen by the human eye, appear to vary in brightness.

For example, in the daylight, human eyes are most sensitive to greenish-yellow light with a wavelength of around 5500 Angstrom units [29].

Furthermore, the eye sees differently from different angles, and color perception decreases with the increase in the viewing angle.

During the design process, careful attention should be given to factors such as these listed below [8,27].

- Do not rely too much on color when critical tasks are to be carried out by fatigued individuals.
- Color makes very little difference in dark surroundings.
- Select colors in such a manner that color-weak persons do not get confused.
- Make use of red filters, if possible, with a wavelength greater than 6500 Angstrom units.
- Ensure that warning lights are as close as possible to red color.

3.10.3 Noise

Noise may be described as a sound that lacks coherence. Human reactions to noise include boredom, irritability, and fatigue. Excessive noise can lead to various types of problems including loss in hearing if exposed for long periods, adverse effects on tasks requiring a high degree of muscular coordination or intense concentration, and reduction in the workers' efficiency.

Two major physical characteristics (i.e., intensity and frequency) are used in describing noise. Intensity is measured in decibels (dB), and a noise level below 90 dB is generally considered to be safe for humans. In the case of frequency, human ears are most sensitive to frequencies in the range of 600–900 Hz. However, humans can suffer a major loss of hearing when they are exposed to noise between 4000 Hz and 6000 Hz for a long time [29,30].

3.10.4 Vibration

The presence of vibrations can be detrimental to the performance of mental and physical tasks by humans. Low-frequency and large-amplitude vibrations contribute to problems such as headaches, motion sickness, eye strain, fatigue, and interference with the ability to read and interpret instruments [29]. These symptoms become less pronounced as the frequency of vibration increases and the amplitude decreases. However, high-frequency and low-amplitude vibrations can also be quite fatiguing.

Some guidelines to reduce the effects of vibration and motion are as follows [29,31]:

- Eliminate or minimize vibrations and shocks through designs and/or using items such as shock absorbers, springs, cushioned mountings, and fluid couplings.

- Eliminate any seating design that would lead to or would transmit 3 to 4 cycles per second vibrations, since the resonant frequency of the human vertical trunk in the seated position is between 3 and 4 cycles per second.
- Aim to eliminate vibrations with amplitudes more than 0.08 mm to perform critical maintenance or other tasks requiring digit or letter discrimination.

3.11 Human Factors-Related Formulas

Professionals working in the area of human factors have developed many mathematical formulas for estimating various types of human factors-related information. Some formulas that are useful for performing patient safety-related analysis are presented below.

3.11.1 Character Height Estimation: Formula I

This formula is concerned with estimating the optimum character height by considering various factors including viewing conditions, viewing distance, and the importance of reading accuracy. The character height is expressed by [32,33]

$$H_c = \theta_1 + \theta_2 + (0.0022)V_d \tag{3.1}$$

where

H_c = the character height expressed in inches.

θ_1 = the correction factor for viewing and illumination conditions. The recommended values for different conditions are 0.06 (above 1 foot-candle, favorable reading conditions), 0.16 (below 1 foot-candle, favorable reading conditions), 0.16 (above 1 foot-candle, unfavorable reading conditions), and 0.26 (below 1 foot-candle, unfavorable reading conditions).

θ_2 = the correction factor for the criticality of the number. For important items such as emergency labels, the recommended value is 0.075, and for other items the recommended value is zero.

V_d = the viewing distance expressed in inches.

Example 3.1

A medical equipment's dials' estimated viewing distance is 30 inches. Calculate the height of the label characters to be used at the dials, if the values of θ_1 and θ_2 are 0.16 and 0.075, respectively.

By inserting the specified values into Equation (3.1), we get

$$H_c = 0.16 + 0.075 + (0.0022)(30)$$
$$= 0.301 \text{ inches}$$

Thus the height of the label characters to be used at the dials is 0.301 inches.

3.11.2 Rest Period Estimation: Formula II

This formula is concerned with estimating the amount of rest (scheduled and unscheduled) required for any specified task or activity. The rest required is expressed by [34]

$$R_r = \frac{WT(\lambda_1 - \lambda_2)}{(\lambda_1 - \alpha)} \tag{3.2}$$

where

R_r = the amount of rest required expressed in minutes.
WT = the amount of working time expressed in minutes.
λ_1 = the average kilocalories expenditure per minute of work.
λ_2 = the level of energy expenditure expressed in kilocalories per minute, adopted as standard.
α = the approximate resting level expressed in kilocalories per minute with its value taken as 1.5.

Example 3.2

A health care worker is performing a task for 120 minutes and his/her average energy expenditure is 5 kilocalories per minute. Calculate the amount of rest required, if the value of λ_2 is 4 kilocalories per minute.

By inserting the specified data values into Equation (3.2), we obtain

$$R_r = \frac{120(5-4)}{5-1.5} = 34.29 \text{ minutes}$$

Thus the amount of rest required (i.e., the length of the required rest period) is 34.29 minutes.

3.11.3 Brightness Contrast Estimation: Formula III

This formula is concerned with estimating brightness contrast. Brightness contrast is expressed by [27]

$$B_c = \frac{(L_b - L_d)(100)}{L_b} \tag{3.3}$$

where

L_d = the luminance of the darker of two contrasting areas.
L_b = the luminance of the brighter of two contrasting areas.
B_c = the brightness contrast.

Example 3.3

Assume that a certain piece of paper has a reflectance of 85%. Estimate the value of the brightness contrast, if the print on the paper has a reflectance of 15%.

By inserting the given data values into Equation (3.3), we obtain

$$B_c = \frac{(85 - 15)(100)}{85}$$
$$= 82.35\%$$

Thus the value of the brightness contrast is 82.35%.

3.11.4 Noise Reduction Estimation: Formula IV

As noise can be a major problem in various facilities including health care, its reduction is essential. This formula is concerned with estimating the reduction. Noise reduction is expressed by [35,36]

$$N_r = \log(APW / WA) + \mu \tag{3.4}$$

where

N_r = the total noise reduction.
APW = the total absorption properties of walls in the noise-receiving room.
WA = the area of wall transmitting sound expressed in ft^2.
μ = the transmission loss of materials of varying thickness expressed in decibels.

Nonetheless, μ is defined by the following equation:

$$\mu = \frac{WA}{(\alpha_1 WA_1 + \alpha_2 WA_2 + \ldots + \alpha_m WA_m)} \tag{3.5}$$

where

α_i = the ith transmission coefficient of material under consideration, for $i = 1, 2, 3, \ldots, m$.
WA_i = the ith corresponding area of the material under consideration.

3.11.5 Glare Constant Estimation: Formula V

This formula is concerned with estimating the value of the glare constant. The glare constant is defined by [33]

$$C_g = [(S_a)^{0.8}(S_l)^{1.6}]/[(L_g)A^2] \tag{3.6}$$

where
S_a = the solid angle subtended by the source at the eye.
L_g = the general background luminance.
S_l = the source luminance.
A = the angle between the viewing direction and the glare source direction.
C_g = the glare constant. It is to be noted that $C_g = 150$ means the boundary of "just uncomfortable" glare, and $C_g = 35$ means the boundary of "just acceptable" glare.

3.11.6 Inspector Performance Estimation: Formula VI

This formula is concerned with estimating inspector performance associated with inspection-related tasks. The inspector performance is expressed by [27,37]

$$INS_p = \frac{RT}{(n-m)} \tag{3.7}$$

where
m = the total number of inspector errors.
n = the total number of patterns inspected.
RT = the total reaction time.
INS_p = the inspector performance expressed in minutes per correct inspection.

3.12 Problems

1. Discuss the safety and human factors historical developments.

2. Discuss the need for safety.
3. List at least eight safety-related facts and figures.
4. What are the common causes of work injuries?
5. Describe the human factors accident causation theory.
6. Describe the following injuries:
   - Shearing injuries
   - Puncturing injuries
   - Breaking injuries
7. Discuss safety management principles.
8. What are the disciplines that contribute to human factors?
9. List at least 10 typical human behaviors.
10. Assume that a health care worker is performing a task for 150 minutes and his/her average energy expenditure is 4 kilocalories per minute. Calculate the length of the rest period required, if the value of λ_2 in Equation (3.2) is 3 kilocalories per minute.

3.13 References

1. Goetsch, D. L., *Occupational Safety and Health*, Prentice Hall, Englewood Cliffs, New Jersey, 1996.
2. Dhillon, B. S., *Engineering Safety: Fundamentals, Techniques, and Applications*, World Scientific Publishing, River Edge, New Jersey, 2003.
3. Heinrich, H. W., *Industrial Accident Prevention*, 3rd Edition, McGraw-Hill, New York, 1950.
4. *Accidental Facts*, Report, National Safety Council, Chicago, Illinois, 1996.
5. Chapanis, A., *Man-Machine Engineering*, Wadsworth, Belmont, California, 1965.
6. Meister, D., Rabideau, G. F., *Human Factors in System Development*, John Wiley and Sons, New York, 1965.
7. Dale Huchingson, R., *New Horizons for Human Factors in Design*, McGraw-Hill, New York, 1981.
8. Dhillon, B. S., *Human Reliability and Error in Transportation Systems*, Springer-Verlag, London, 2007.
9. Dhillon, B. S., *Reliability, Quality, and Safety for Engineers*, CRC Press, Boca Raton, Florida, 2005.
10. Report No. OSHA 2056, *All about OSHA*, U.S. Department of Labor, Washington, D.C., 1991.
11. *Report on Injuries in America in 2000*, National Safety Council, Chicago, Illinois, 2000.
12. Hammer, W., Price, D., *Occupational Safety Management and Engineering*, Prentice Hall, Upper Saddle River, New Jersey, 2001.

13. *How to Reduce Workplace Accidents*, Report, European Agency for Safety and Health at Work, Brussels, Belgium, 2001.
14. *Accident Facts*, National Safety Council, Chicago, Illinois, 1990–1993.
15. Banta, H. D., The Regulation of Medical Devices, *Preventive Medicine*, Vol. 19, 1990, pp. 693–699.
16. *Medical Devices*, Hearings Before the Sub-Committee on Public Health and Environment, U.S. Congress Interstate and Foreign Commerce, Serial No. 93-61, U.S. Government Printing Office, Washington, D.C., 1973.
17. Schneider, P., Hines, M. L. A., Classification of Medical Software, *Proceedings of the IEEE Symposium on Applied Computing*, 1990, pp. 20–27.
18. Gowen, L. D., Yap, M. Y., Traditional Software Development's Effects on Safety, *Proceedings of the 6th Annual IEEE Symposium on Computer-Based Medical Systems*, 1993, pp. 58–63.
19. Hunter, T. A., *Engineering Design for Safety*, McGraw-Hill, New York, 1992.
20. Heinrich, H. W., Peterson, D., Roos, N., *Industrial Accident Prevention*, McGraw-Hill, New York, 1980.
21. Petersen, D., *Techniques of Safety Management*, McGraw-Hill, New York, 1971.
22. Petersen, D., *Safety Management*, American Society of Safety Engineers, Des Plaines, Illinois, 1998.
23. Chapanis, A., *Human Factors in Systems Engineering*, John Wiley and Sons, New York, 1996.
24. Dhillon, B. S., *Engineering Design: A Modern Approach*, Richard D. Irwin, Chicago, 1996.
25. Woodson, W. E., *Human Factors Design Handbook*, McGraw-Hill, New York, 1981.
26. *Anthropometry for Designers, Anthropometric Source Book I*, Report No. 1024, National Aeronautics and Space Administration, Houston, Texas, 1978.
27. Dhillon, B. S., *Human Reliability with Human Factors*, Pergamon Press, New York, 1986.
28. Lederman, S., Heightening Tactile Impression of Surface Texture, in *Active Touch*, edited by G. Gordon, Pergamon Press, New York, 1978, pp. 40–45.
29. AMCP 706-134, *Engineering Design Handbook: Maintainability Guide for Design*, Prepared by the United States Army Material Command, 5001 Eisenhower Avenue, Alexandria, Virginia, 1978.
30. AMCP 706-133, *Engineering Design Handbook: Maintainability Engineering Theory and Practice*, prepared by the United States Army Material Command, 5001 Eisenhower Avenue, Alexandria, Virginia, 1976.
31. Altman, J. W., et al., *Guide to Design to Mechanical Equipment for Maintainability*, Report No. ASD-TR-61-381, Air Force Systems Command, Wright-Patterson Air Force Base, Ohio, 1961.
32. Peters, G. A., Adams, B. B., Three Criteria for Readable Panel Markings, *Product Engineering*, Vol. 30, 1959, pp. 55–57.
33. Osborne, D. J., *Ergonomics at Work*, John Wiley and Sons, New York, 1982.
34. Murrell, K. F. H., *Human Performance in Industry*, Reinhold, New York, 1965.
35. Dale Huchingson, R., *New Horizons for Human Factors in Design*, McGraw-Hill, New York, 1981.
36. Dhillon, B. S., *Human Reliability and Error in Medical System*, World Scientific Publishing, River Edge, New Jersey, 2003.
37. Drury, C. G., Fox, J. G., Editors, *Human Reliability in Quality Control*, John Wiley and Sons, New York, 1975.

4

Methods for Performing Patient Safety Analysis

4.1 Introduction

Today, the fields of safety, human factors, quality, and reliability are well-established disciplines. A large number of publications in the form of journal articles, conference proceedings articles, technical reports, and books on each of these disciplines have appeared. These publications include many new concepts and methods. Some of the methods developed in these fields are being used across many diverse disciplines including engineering design, manufacturing, and management.

One example of these methods is fault tree analysis (FTA), developed in the area of reliability engineering. FTA is used to perform design, manufacturing, maintenance, and management studies. Another example of a widely used method across many diverse disciplines is called the failure modes and effect analysis (FMEA). This method was also originally developed for applications in the field of reliability engineering.

Many methods and approaches that have been developed in the fields of reliability, safety, quality, and human factors are being used in various other areas. This chapter presents a number of such methods and approaches used to perform patient safety analysis.

4.2 Failure Modes and Effect Analysis (FMEA)

FMEA is a method widely used in the industrial sector to perform reliability and safety analyses of engineering systems. It is a powerful tool used to perform analysis of each potential failure mode in a system to determine the effects of such failure modes on the total system [1,2]. When FMEA is extended to classify the effect of each potential failure according to its severity, it is called failure mode effects and criticality analysis (FMECA).

The method was developed in the early 1950s by the Bureau of Aeronautics of the U.S. Navy and was called *failure analysis* [3]. Subsequently, the technique was renamed *failure effect analysis* and the Bureau of Naval Weapons inducted it into its new specification on flight control systems [4]. The National Aeronautics and Space Administration (NASA) extended the FMEA's functions and renamed it FMECA [5]. In the 1970s, the U.S. Department of Defense developed a military standard called "Procedures for Performing a Failure Mode, Effects, and Criticality Analysis" [6].

The main steps followed in performing FMEA are shown in Figure 4.1 [5,7]. Step 1 is basically concerned with breaking down the system into blocks, block functions, and the interface between them. Usually at the initial stages of a program, a reasonably good system definition does not exist and the analyst develops his/her own system definition by using documents such as development plans and specifications, trade study reports, and drawings.

Step 2 is concerned with developing ground rules as to how the FMEA will be subsequently conducted. After the completion of the system definition and the mission requirements, the development of ground rules is a straightforward process. Some examples of the ground rules are definition of what

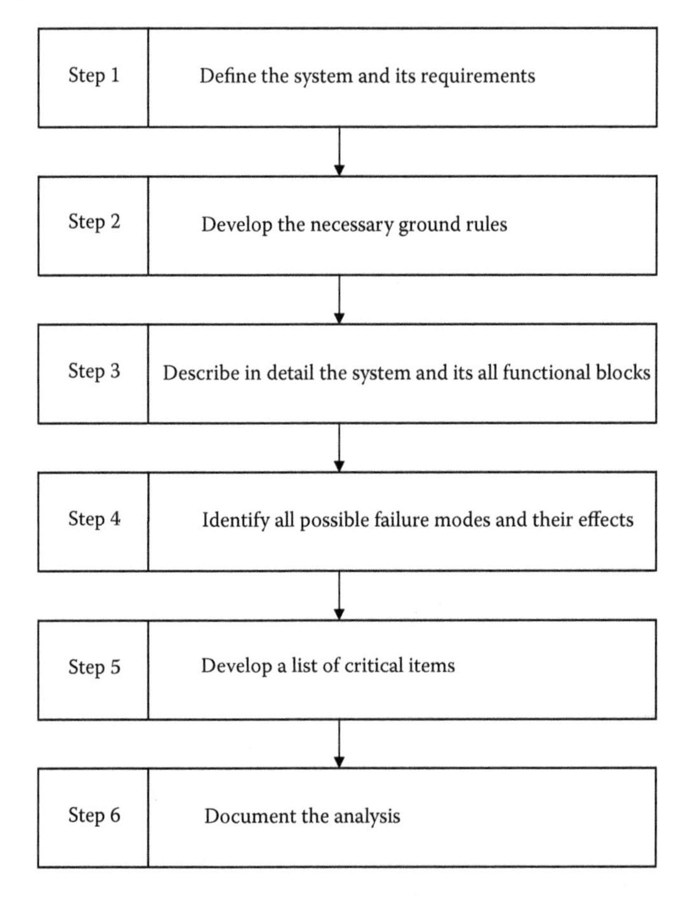

FIGURE 4.1
Main steps involved in performing FMEA.

constitutes failure of system hardware components, limits of environmental and operational stresses, delineation of mission phases, and description of the coding system used.

Step 3 is concerned with preparing the system description, which can be grouped under two parts: system block diagram and narrative functional statement. The system block diagram is used to determine the success/failure relationships among all the system parts. The narrative functional statement provides a narrative description of each item's operation for each operational mode/mission phase. Step 4 is concerned with performing analysis of the failure modes and their effects. In this regard, usually a well-designed form or a worksheet is used. The form collects information on many areas including item identification and function, failure modes and causes, effects of failures on system mission/personnel/subsystems, failure detection method, criticality classification, compensating provisions, and other remarks.

Step 5 is concerned with developing a list of critical items. The list is prepared to facilitate the communication of important analysis results, generally to the management personnel, and it contains information on areas such as item identification, concise statement of failure mode, criticality classification, and degree of loss effect. Finally, Step 6 is concerned with documenting the analysis. This step is equivalent in importance to all the previous steps (i.e., Step 1 to Step 5) because poor documentation can result in ineffectiveness of the FMEA process. The FMEA report includes items such as system description, ground rules, system definition, failure modes and their effects, and critical items list.

Professionals working in the area of reliability and safety analysis have established certain facts/guidelines concerning FMEA, some of which follow [8]:

- FMEA has certain limitations.
- FMEA is not a method for choosing the optimum design concept.
- FMEA is not designed to supersede the engineer's work.
- Avoid developing the majority of the FMEA in a meeting.

4.2.1 FMEA Advantages

FMEA has been successfully applied in many diverse areas. Some of its advantages are as follows [8,9]:

- It provides an effective safeguard against making the same mistakes in the future.
- It is easy to understand and helps to improve customer satisfaction.
- It identifies safety-related concerns to be focused on.
- It helps to reduce the item development time and cost.

- It provides a visibility tool for management.
- It helps to improve communication among design interface personnel.
- It is a useful method that begins from the detailed level and works upward.
- It is a useful tool to compare alternative designs.
- It is a systematic approach to classify hardware failures.

The applications of FMEA in the area of health care are available in Refs. [10–12].

4.3 Root Cause Analysis (RCA)

This method has been used for many years in the industrial sector to investigate industrial incidents and was originally developed by the U.S. Department of Energy [13,14]. RCA may be described as a systematic investigation approach that uses information collected during an assessment of an accident to determine the underlying factors for the deficiencies or shortcomings that led to the occurrence of the accident [15].

RCA begins with outlining the event sequence leading to the occurrence of the accident under consideration. Starting with the adverse event itself, the analyst performs his/her functions backward in time, recording and ascertaining all important events. On the basis of investigational findings, the RCA process concludes with appropriate recommendations for improvements.

Ten general steps to perform RCA in the area of health care are as follows [1,16]:

- **Step 1:** Educate all concerned people about RCA.
- **Step 2:** Inform all staff personnel whenever a sentinel event is reported.
- **Step 3:** Form an RCA team of appropriate individuals.
- **Step 4:** Prepare for and conduct the first team meeting.
- **Step 5:** Determine the sequence of the event.
- **Step 6:** Identify and separate each event sequence that may have been a contributory factor in the sentinel event occurrence.
- **Step 7:** Brainstorm about the factors surrounding the selected events that may have been contributory to the sentinel event occurrence.
- **Step 8:** Affinitize with the brainstorming session results.
- **Step 9:** Develop an appropriate action plan.

- **Step 10:** Distribute the RCA document and the associated action plan to all concerned individuals.

The Joint Commission on the Accreditation of Healthcare Organizations (JCAHO) in the United States recommends that all health care facilities in the country respond to sentinel events by using RCA within 45 days of their occurrence.

There are many software packages available in the market to perform RCA. Some of these are as follows [15]:

- **TAPROOT:** System Improvements, Inc., 238 S. Peters Road, Suite 301, Knoxville, TN 37923-5224.
- **REASON:** Decision Systems, Inc., 802 N. High St., Suite C, Longview, TX 75601.
- **BRAVO:** JBF Associates, Inc., 1000 Technology Drive, Knoxville, TN 37939.

The applications of RCA in the area of health care are available in Refs. [17,18].

4.3.1 RCA Advantages and Disadvantages

Professionals working in the area of safety have identified many advantages and disadvantages of RCA. Some of its advantages are as follows [1,19]:

- It is an effective tool to identify and address organizational and system-related issues.
- The systematic application of RCA can help to uncover common root causes that link a disparate collection of accidents.
- It is a well-structured and process-focused method.

Some of the disadvantages of RCA are as follows [1,19]:

- It is labor intensive and time consuming.
- It is not possible to determine exactly if the root cause established by the analysis is the real cause of the accident.
- RCA is essentially an uncontrolled case study.
- RCA can be tainted by hindsight bias.

4.4 Hazard Operability Analysis (HAZOP)

HAZOP is a systematic and effective approach used to identify hazards and operating problems in a facility. It was originally developed in the chemical

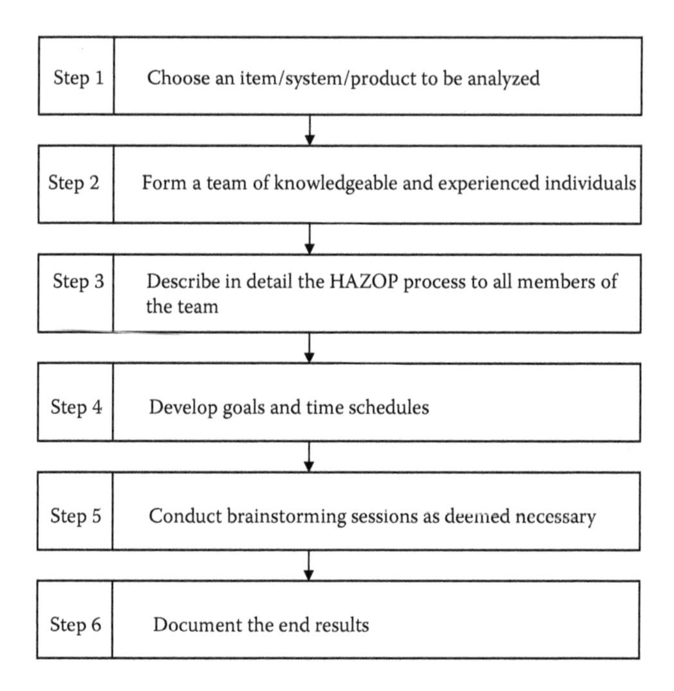

FIGURE 4.2
HAZOP study steps.

industry to perform safety studies. The method is useful for highlighting unforeseen potential hazards designed into facilities due to various reasons, or introduced into existing facilities due to factors such as changes made to process conditions or operating procedures.

The method calls for forming a team composed of experienced and knowledgeable members with different backgrounds. These team members brainstorm about possible potential hazards. The team is led by a highly experienced person and during the brainstorming sessions, the same person acts as facilitator.

A HAZOP study is conducted in six steps, as shown in Figure 4.2 [20]. One important drawback of the HAZOP is that it does not take into account the human error occurrence in the final equation. Additional information on this method is available in Ref. [21].

4.5 Interface Safety Analysis (ISA)

This method is used to determine incompatibilities between subsystems and assemblies of an item/system/product that could result in accidents. The method establishes that distinct parts/units can be integrated into a viable

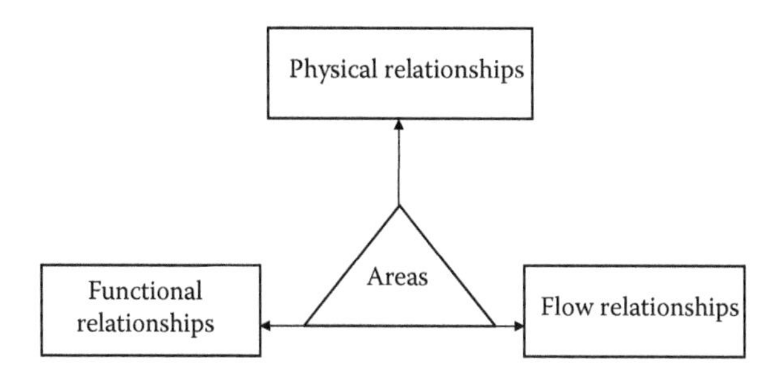

FIGURE 4.3
ISA relationships grouping areas.

system and an individual part or unit normal operation will not impair the performance or damage another part/unit or the entire system/product.

ISA considers various relationships, but they can be grouped basically into three areas as shown in Figure 4.3 [22,23]. The *functional relationships* are concerned with multiple units or items. For example, in a situation where outputs of a single unit constitute the inputs to a downstream item or unit, any error in outputs and inputs may cause damage to the downstream item or unit, and in turn a serious safety problem or hazard. The outputs could be in conditions such as listed below:

- Erratic outputs
- Zero outputs
- Degraded outputs
- Excessive outputs
- Unprogrammed outputs

The *flow relationships* are concerned with two or more units or items. For example, the flow between two units may involve air, water, steam, lubricating oil, fuel, or electrical energy. In addition, the flow could be unconfined, such as heat radiation from one body (unit) to another body (unit). Problems associated with many products are the proper flow of energy or fluids from one item or unit to another item or unit through confined spaces or passages, consequently leading to safety-related problems. The flow-related problem causes include faulty connections between units and complete or partial interconnection failure. In the case of fluid, from the safety aspect factors such as those listed below must be considered carefully:

- Flammability
- Odor
- Toxicity
- Lubricity

- Corrosiveness
- Contamination
- Loss of pressure

Finally, the *physical relationships* are concerned with the physical aspects of items or units. For example, two items/units might be very well designed and manufactured and operate well individually, but they may fail to fit together properly due to dimensional differences or they may represent other problems that may lead to safety-related problems. Three examples of the other problems are as follows:

- A very small clearance between units; thus the units may be damaged during the removal process.
- Impossible or restricted access to or egress from equipment.
- Impossible to tighten, join, or mate parts properly.

4.6 Preliminary Hazard Analysis (PHA)

This approach is generally used during the conceptual design phase and is relatively an unstructured tool because of the lack or unavailability of definitive information such as functional flow diagrams and drawings [23]. PHA has proven to be an effective method to take appropriate early steps to identify and eliminate/minimize hazards when the required data are not available. The findings of PHA serve as an effective guide in potential detailed analysis.

The method calls for the formation of an ad hoc team made up of members familiar with items such as equipment, material, substance, and/or process under consideration. To perform an effective PHA, the appropriate experience and related expertise of all team members is essential. All members of the team are required to review the occurrence of all types of hazards in the area of their expertise, and as a group or team they play the devil's advocate.

Additional information on PHA is available in Refs. [24,25].

4.7 Technic of Operations Review (TOR)

This method was developed in the early 1970s by D. A. Weaver of the American Society of Safety Engineers to identify systemic causes rather than assigning blame in regard to safety [26]. TOR may be described as a hands-on analytical methodology developed to determine the root system causes of an operation failure [27]. The method makes use of a worksheet containing simple terms requiring yes/no decisions.

The basis for the activation of this approach is an incident occurring at a certain place and time involving certain individuals. The following eight steps are used in this approach [26,27]:

- **Step 1:** Form the TOR team by choosing its members from all the appropriate areas.
- **Step 2:** Hold a roundtable session to communicate common knowledge to all team members.
- **Step 3:** Identify one important systemic factor that has played a key role in the occurrence of accident/incident under consideration. This factor is the result of team consensus and serves as an initial point for all further investigations.
- **Step 4:** Use the team consensus to respond to a sequence of yes/no options.
- **Step 5:** Evaluate the identified factors and ensure the existence of consensus among the team members with respect to the evaluation of each factor.
- **Step 6:** Prioritize all the contributing factors starting with the most serious.
- **Step 7:** Develop appropriate preventive/corrective strategies in regard to each contributing factor.
- **Step 8:** Implement strategies.

Just like any other safety analysis method, the TOR has its strengths and weaknesses. Its main strength is the involvement of line personnel/people in the analysis, and its main weakness is its after-the-fact process.

4.8 Job Safety Analysis (JSA)

This method is used to find and rectify potential hazards that are intrinsic to or inherent in a workplace. Normally, people such as the safety professional, supervisor, and worker participate in job safety analysis. The following five steps are associated with JSA [28]:

- **Step 1:** Choose a job for analysis.
- **Step 2:** Break down the job under consideration into a number of tasks or steps.

- **Step 3:** Identify all types of potential hazards and determine suitable measure to control them effectively.
- **Step 4:** Apply the determined measures.
- **Step 5:** Evaluate the final results.

The degree of JSA success depends on the thoroughness exercised by the JSA team members during the analysis process.

4.9 Six Sigma Methodology

This methodology was originally developed by Motorola USA in 1986 as a business management strategy [29]. It involves designing, monitoring, and improving process to eliminate or minimize unnecessary waste while optimizing general satisfaction and increasing financial-related stability [29,30]. The process performance is used to measure improvement by comparing the baseline process capability (i.e., prior to improvement) with the process capability subsequent to piloting all possible potential solutions for improving quality [31,32].

The following two basic methods are used with the Six Sigma methodology:

- **Method I:** This method inspects process and counts all types of defects, calculates a defect rate per million, and makes use of a statistical table for converting defect rate per million to a sigma (σ) metric. The method is applicable to postanalytic and preanalytic processes.
- **Method II:** This method makes use of estimates of process variation for predicting process performance by estimating a sigma (σ) metric from the specified tolerance limits and the variations observed for the process under consideration. The method is useful for analytic processes in which the accuracy and precision can be determined through experimental procedures.

One element of Six Sigma methodology makes use of a five-phased process; define, measure, analyze, improve, and control (DMAIC) approach [33,34]. To begin the methodology, the project is identified, the scope of expectations is defined, and historical data are reviewed. In the next step, continuous total quality performance standards are chosen and performance objectives defined along with sources of variability. At the implementation of the new project, appropriate data are collected for assessing how well changes have helped to improve the process. To support this type of analysis, appropriate validated measures are developed for determining the capability of the new process.

Additional information on the application of Six Sigma in the area of health care is available in Refs. [31–34].

4.10 Pareto Diagram

This is a simple bar chart that ranks related problems/measures in decreasing frequency of occurrence and is often used for selection of a limited number of tasks that generate significant overall effect. The diagram is named after an Italian sociologist and economist, Vilfredo Pareto (1848–1923), who in the early 1900s conducted a study on the spread of poverty and wealth in Europe. He was surprised to learn that wealth in Europe was concentrated in the hands of about 20% of the people while poverty afflicted around 80% of the people [35,36].

Pareto's findings may be referred to as the law of the "significant few versus the trivial many." The Pareto diagram approach helps to identify the top 20% of causes that need to be addressed to resolve 80% of the problems. In other words, the objective of the Pareto diagram is to separate the significant aspects of a problem under consideration from the trivial ones. This in turn helps decision makers determine where to direct improvement-related efforts.

The following eight steps are associated with the construction of a Pareto diagram [36,37]:

- **Step 1:** Determine the method to classify the data (e.g., by type of nonconformity, problem, cause).
- **Step 2:** Decide what is to be used for ranking the characteristics (e.g., frequency or dollars).
- **Step 3:** Collect the required data (e.g., for an appropriate time interval).
- **Step 4:** Summarize the collected data.
- **Step 5:** Rank classifications (e.g., from largest to smallest).
- **Step 6:** Calculate the cumulative percentage, if necessary.
- **Step 7:** Develop/construct Pareto diagram.
- **Step 8:** Identify the vital few.

4.11 Fault Tree Analysis (FTA)

This is a widely used method in the industrial sector, particularly in nuclear power generation, to perform reliability and safety analysis of engineering systems during the design and development phase. The method was developed in the early 1960s at the Bell Telephone Laboratories in the United States to perform reliability analysis of the Minuteman Launch Control System [9].

A fault tree is a logical representation of the relationship or primary events that may cause the occurrence of a specified undesirable event, known as the *top event*, and it is depicted using a tree structure with normally OR and AND logic gates. The method is described in detail in Ref. [9], and an extensive list of publications on the method is available in Ref. [38].

4.11.1 Common Fault Tree Symbols and Fault Tree Analysis Steps

Although many symbols are used in the construction of fault trees, the four most commonly used symbols are shown in Figure 4.4. The OR gate means that an output fault event occurs if one or more input fault events occur. The AND gate means that an output fault event occurs only if all the input fault events occur. The circle denotes a basic fault event or the failure of an elementary part. The values of this fault event's parameters such as the occurrence probability, the occurrence failure rate, and the repair rate are usually obtained from the empirical data. Finally, the rectangle represents a fault event that results from the logical combination of fault events through the input of a logic gate such as OR and AND.

The following seven steps are generally used to perform fault tree analysis [1,39]:

- **Step 1:** Define the system and all its associated assumptions.

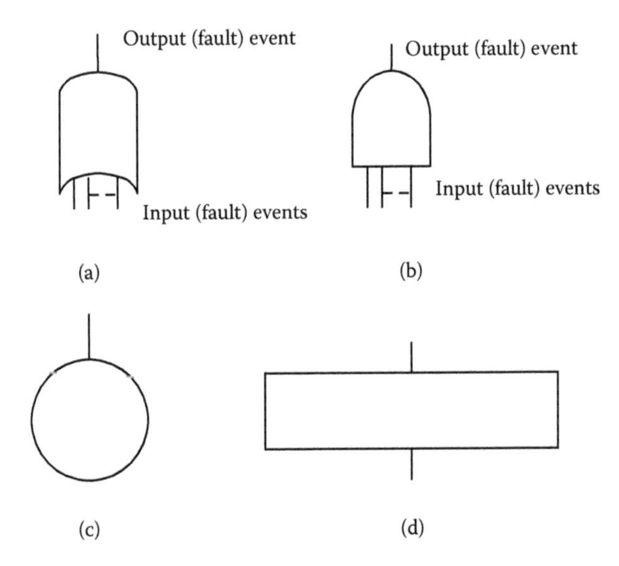

FIGURE 4.4
Basic fault tree symbols: (a) OR gate, (b) AND gate, (c) circle, (d) rectangle.

- **Step 2:** Identify the undesirable event (i.e., the top fault event) to be investigated (e.g., system failure).
- **Step 3:** Determine all types of possible potential causes that can result in the occurrence of the top fault event by using fault tree symbols, such as shown in Figure 4.4, and the logic tree format.
- **Step 4:** Develop the fault tree to the lowest level of detail as per the specified requirements.
- **Step 5:** Perform analysis of the completed fault tree in regard to factors such as gaining insight into the unique modes of item/product faults and understanding the appropriate logic and the interrelationships among the fault paths.
- **Step 6:** Determine the necessary corrective measures.
- **Step 7:** Document the analysis and follow up on the corrective measures.

Example 4.1

Assume that a patient in a health care facility receives a medication from a nursing professional prescribed by a doctor. The patient can be given incorrect medication or the wrong amount due to either doctor error or nursing error. The doctor error can occur due to misdiagnosis, haste, or poor surroundings. Similarly, the nursing error can occur due to wrong interpretation of the doctor's instructions, haste, or poor work environment.

Wrong interpretation of the doctor's instructions can occur due to either poor verbal instructions or poorly written instructions.

With the aid of fault tree symbols given in Figure 4.4, develop a fault tree for the top event "patient given incorrect medication or wrong amount."

A fault tree for the example is shown in Figure 4.5. The single capital letters in the figure denote corresponding fault events (e.g., Y: wrong interpretation of doctor's instructions, B: misdiagnosis, and E: poor verbal instructions).

4.11.2 Probability Evaluation of Fault Trees

In the event that the probability of the occurrence of basic fault events is given, it is possible to calculate the probability of occurrence of the top fault event. This can only be calculated by first calculating the probability of occurrence of the output fault events of the intermediate and lower logic gates such as the OR and the AND gates.

The probability of occurrence of the OR gate output fault event is expressed by [9]

$$P(Y) = 1 - \prod_{j=1}^{k} \{1 - P(Y_j)\} \tag{4.1}$$

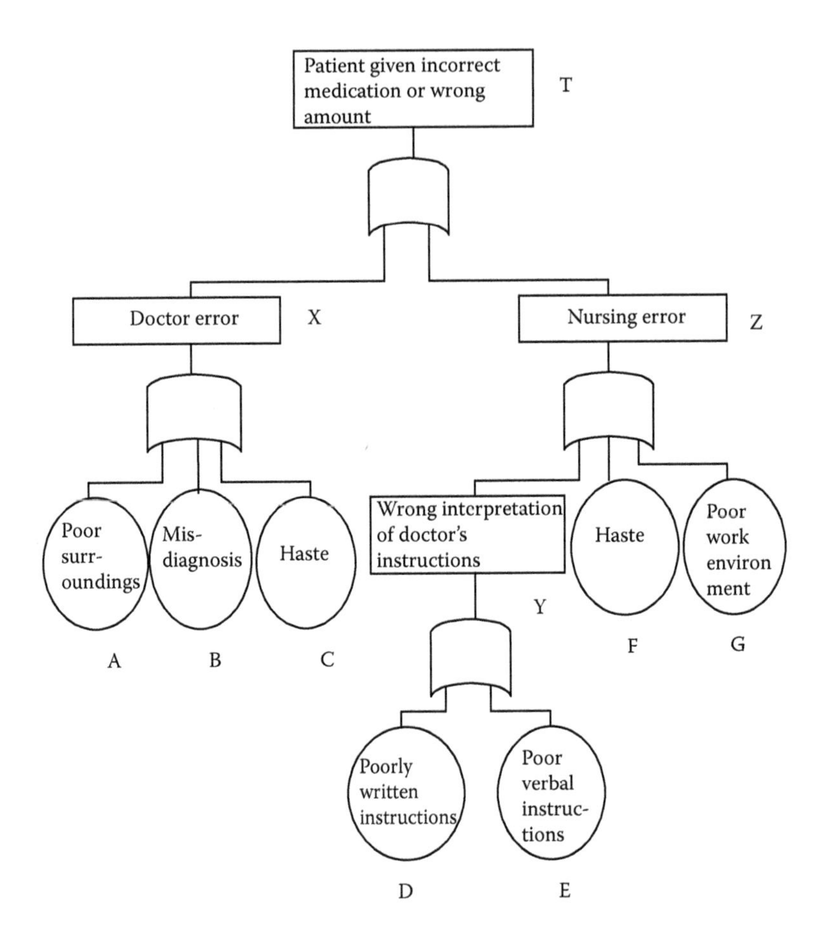

FIGURE 4.5
A fault tree for Example 4.1.

where

k = the number of input fault events.

$P(Y_j)$ = the probability of occurrence of OR gate input fault event Y_j, for j = 1, 2, 3, ..., k.

$P(Y)$ = the probability of occurrence of the OR gate output fault event Y.

Similarly, the probability of occurrence of the AND gate output fault event is expressed by [9]

$$P(X) = \prod_{j=1}^{k} P(X_j) \tag{4.2}$$

where

k = the number of input fault events.
$P(X_j)$ = the probability of occurrence of AND gate input fault event X_j, for j = 1, 2, 3, ..., k.
$P(X)$ = the probability of occurrence of the AND gate output fault event X.

Example 4.2

Assume that in Figure 4.5 the occurrence probabilities of fault events *A*, *B*, *C*, *D*, *E*, *F*, and *G* are 0.07, 0.06, 0.05, 0.04, 0.03, 0.02, and 0.01, respectively. Calculate the probability of occurrence of the top fault event *T*, patient given incorrect medication or wrong amount, by using Equation (4.1).

By substituting the given occurrence probability values of fault events *A*, *B*, and *C* into Equation (4.1), the probability of the doctor error, *X*, is

$$P(X) = 1 - (1 - 0.07)(1 - 0.06)(-0.05)$$
$$= 0.1696$$

Similarly, by substituting the specified occurrence probability values of fault events *D* and *E* into Equation (4.1), the probability of the wrong interpretation of doctor's instructions, *Y*, is

$$P(Y) = 1 - (1 - 0.04)(1 - 0.03)$$
$$= 0.0688$$

By inserting the given occurrence probability values of fault events *F* and *G* and the calculated value of fault event *Y* into Equation (4.1), the probability of the nursing error, *Z*, is

$$P(Z) = 1 - (1 - 0.02)(1 - 0.01)(1 - 0.0688)$$
$$= 0.0965$$

Substituting the above calculated values for fault events *X* and *Z* into Equation (4.1), we get

$$P(T) = 1 - (1 - 0.1695)(1 - 0.0965)$$
$$= 0.2496$$

where
$P(T)$ = the probability of occurrence of the top fault event, *T*.

Thus, the probability of the patient being given incorrect medication or wrong amount is 0.2496. Figure 4.6 shows the Figure 4.5 fault tree with the above calculated and specified fault event occurrence probability values.

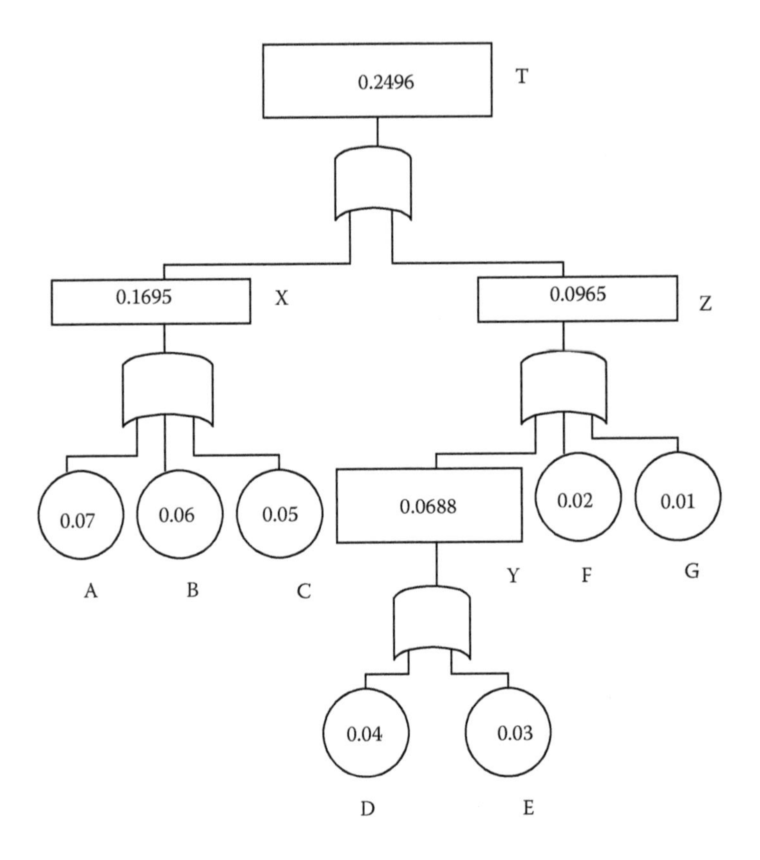

FIGURE 4.6
A fault tree with given and calculated fault event occurrence probability values.

4.11.3 Fault Tree Analysis Benefits and Drawbacks

Over the years professionals working in the area of reliability and safety engineering have identified many benefits and drawbacks of fault tree analysis. Some of its main benefits are as follows [9,23]:

- Identifies faults/failures deductively
- Allows concentration on one particular fault/failure at a time
- Provides insight into the system behavior
- Requires the analyst to understand thoroughly the system/problem under consideration prior to starting the analysis
- Handles complex systems more easily
- Serves as a graphic aid for management personnel
- Provides options for management personnel and others to carry out either qualitative or quantitative reliability/safety analysis

Some of the main drawbacks of fault tree analysis are as follows [9,23]:

- It is a costly and time-consuming method.
- The end results are difficult to check.
- It considers components/elements as either working or failed, so the components'/elements' partial failure states are difficult to handle.

4.12 Markov Method

This is a widely used method to perform various types of reliability and safety analysis of engineering systems in the industrial sector. The method is named after Andrei A. Markov (1856–1922), a Russian mathematician. The method can also be used to perform patient safety-related analysis.

The following assumptions are associated with this method [9]:

- All occurrences are independent of each other.
- The probability of the occurrence of a transition from one system state to another in the finite time interval Δt is given by $\lambda \Delta t$, where λ is the transition rate from one system state to another.
- The probability of more than one transition occurrence in the finite time interval Δt from one state to another is negligible (e.g., $(\lambda \Delta t)(\lambda \Delta t) \rightarrow 0$).

The application of the Markov method to a patient safety-related problem is demonstrated through the following example.

Example 4.3

Assume that the nursing professionals at a health care facility make errors at a constant rate λ. A state space diagram, shown in Figure 4.7, exhibits this scenario. The numerals in the figure denote system states. With the aid of the Markov method, develop expressions for calculating the nursing professionals'

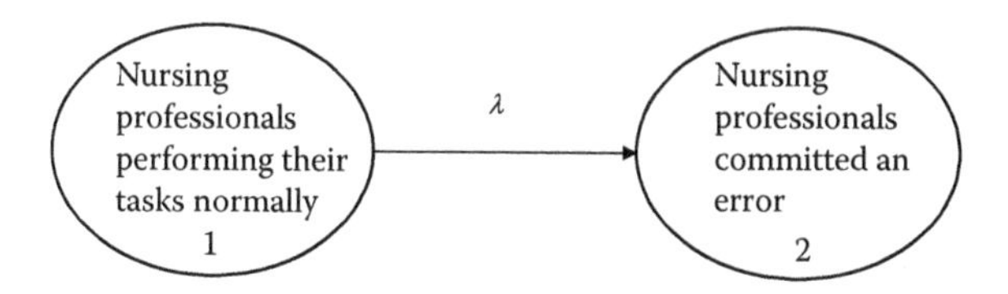

FIGURE 4.7
State space diagram representing the nursing professionals.

reliability and unreliability at time t—in other words, the probability of the nursing professionals performing their tasks without error at time t and the probability of the nursing professionals committing an error at time t, respectively.

Using the Markov method, we write down the following equations for states 1 and 2, respectively, shown in Figure 4.7 [1,9]:

$$P_1(t+\Delta t) = P_1(t)(1-\lambda\Delta t) \tag{4.3}$$

$$P_2(t+\Delta t) = P_2(t) + P_1(t)\lambda\Delta t \tag{4.4}$$

where

- t = time.
- λ = the constant error rate of the nursing professionals.
- $\lambda\Delta t$ = the probability of human error by the nursing professionals in finite time interval Δt.
- $(1 - \lambda\Delta t)$ = the probability of zero human error by the nursing professionals in finite time interval Δt.
- $P_1(t)$ = the probability that the nursing professionals are performing their tasks normally (i.e., state 1 in Figure 4.7) at time t.
- $P_1(t + \Delta t)$ = the probability that the nursing professionals are performing their tasks normally (i.e., state 1 in Figure 4.7) at time $(t + \Delta t)$.
- $P_2(t)$ = the probability that the nursing professionals have committed an error (i.e., state 2 in Figure 4.7) at time t.
- $P_2(t + \Delta t)$ = the probability that the nursing professionals have committed an error (i.e., state 2 in Figure 4.7) at time $(t + \Delta t)$.

By rearranging Equations (4.3) and (4.4), we write

$$\lim_{\Delta t \to o} \frac{P_2(t+\Delta t) - P_1(t)}{\Delta t} = -\lambda P_1(t) \tag{4.5}$$

$$\lim_{\Delta t \to 0} \frac{P_2(t+\Delta t) - P_2(t)}{\Delta t} = \lambda P_1(t) \tag{4.6}$$

Thus, from Equations (4.5) and (4.6), we obtain

$$\frac{dP_1(t)}{dt} + \lambda P_1(t) = 0 \tag{4.7}$$

$$\frac{dP_2(t)}{dt} - \lambda P_1(t) = 0 \tag{4.8}$$

At time $t = 0$, $P_1(0) = 1$ and $P_2(0) = 0$.

By solving Equations (4.7) and (4.8) using Laplace transforms, we obtain

$$P_1(s) = \frac{1}{s+\lambda} \tag{4.9}$$

$$P_2(s) = \frac{\lambda}{s(s+\lambda)} \tag{4.10}$$

where
s = the Laplace transform variable.

By taking the inverse Laplace transforms of Equations (4.9) and (4.10), we get

$$P_1(t) = e^{-\lambda t} \tag{4.11}$$

$$P_2(t) = 1 - e^{-\lambda t} \tag{4.12}$$

Thus the nursing professionals' reliability at time t (i.e., the probability of the nursing professionals performing their tasks without error at time t) and unreliability at time t (i.e., the probability of the nursing professionals committing an error at time t) are given by Equations (4.11), and (4.12), respectively.

Example 4.4

Assume that the constant error rate of nursing professionals in performing their tasks is 0.005 errors per hour. Calculate the reliability and unreliability of the nursing professionals during an 8-hour work period.

By substituting the specified data values (i.e., λ = 0.005 and t = 8) into Equations (4.11) and (4.12), we obtain

$$P_1(8) = e^{-(0.005)(8)}$$
$$= 0.9608$$

and

$$P_2(8) = 1 - e^{-(0.005)(8)}$$
$$= 0.0392$$

Thus the nursing professionals' reliability and unreliability during the specified work period are 0.9608 and 0.0392, respectively.

4.13 Problems

1. Write an essay on methods for performing patient safety analysis.
2. Discuss the steps involved in performing failure modes and effect analysis (FMEA).
3. What are the main advantages of FMEA?
4. Describe root cause analysis (RCA) along with its advantages and disadvantages.
5. Compare hazard operability analysis (HAZOP) with preliminary hazard analysis (PHA).
6. Discuss interface safety analysis (ISA).
7. Describe the following two methods:
   - Technic of operations review (TOR)
   - Job safety analysis (JSA)
8. What is a Pareto diagram?
9. Define the four commonly used symbols to perform fault tree analysis. What are the main advantages and disadvantages of fault tree analysis?
10. Assume that the error rate of a health care professional in performing his/her assigned task is 0.002 errors per hour. Calculate his/her probability of making an error during a 10-hour work period.

4.14 References

1. Dhillon, B. S., *Human Reliability and Error in Medical System*, World Scientific Publishing, River Edge, New Jersey, 2003.
2. Omdahl, T. P., Editor, *Reliability, Availability, and Maintainability (RAM) Dictionary*, American Society for Quality Control (ASQC) Press, Milwaukee, Wisconsin, 1988.
3. MIL-F-18372 (Aer.), *General Specification for Design, Installation, and Test of Aircraft Flight Control Systems*, Bureau of Naval Weapons, Department of the Naval Weapons, Department of the Navy, Washington, D.C., Paragraph 3.5.2.3.
4. Continho, J. S., Failure Effect Analysis, *Transactions of the New York Academy of Sciences*, Vol. 26, Series II, 1963–1964, pp. 564–584.
5. Jordan, W. E., Failure Modes, Effects, and Criticality Analyses, *Proceedings of the Annual Reliability and Maintainability Symposium*, 1972, pp. 30–37.
6. MIL-STD-1629, *Procedures for Performing a Failure Mode, Effects, and Criticality Analysis*, Department of Defense, Washington, D.C., 1980.
7. Dhillon, B. S., *Systems Reliability, Maintainability, and Management*, Petrocelli Books, New York, 1983.

8. Palady, P., *Failure Modes and Effects Analysis*, PT Publications, West Palm Beach, Florida, 1995.
9. Dhillon, B. S., *Design Reliability: Fundamentals and Applications*, CRC Press, Boca Raton, Florida, 1999.
10. Esmail, R., Cummings, C., Dersch, D., et al., Using Healthcare Failure Mode and Effect Analysis Tool to Review the Process of Ordering and Administering Potassium Chloride and Potassium Phosphate, *Healthcare Quality*, Vol. 8, 2005, pp. 73–80.
11. Burgmeier, J., Failure Mode and Effect Analysis: An Application in Reducing Risk in Blood Transfusion, *Jt. Comm. J. Qual. Improv.*, Vol. 28, No. 6, 2002, pp. 331–339.
12. Nickerson, T., Jenkins, M., Greenall, J., Using ISMP Canada's Framework for Failure Mode and Effects Analysis: A Tale of Two FMEAs, *Healthcare Quarterly*, Vol. 11 (Special issue), 2008, pp. 40–46.
13. Busse, D. K., Wright, D. J., *Classification and Analysis of Incidents in Complex, Medical Environments*, Report, 2000. Available from the Intensive Care Unit, Western General Hospital, Edinburgh, U.K.
14. DOE-NE-STD-1004-92, *Root Cause Analysis Guidance Document*, United States Department of Energy (DOE), Washington, D.C., February 1992.
15. Latino, R. J., *Automating Root Cause Analysis, in Error Reduction in Health Care*, edited by P.L. Spath, John Wiley and Sons, New York, 2000, pp. 155–164.
16. Burke, A., *Root Cause Analysis*, Report, 2002. Available from the Wild Iris Medical Association, P.O. Box 257, Comptche, California.
17. Bagian, J. P., Gosbee, J., Lee, C. Z., et al., The Veterans Affairs Root Cause Analysis System in Action, *Jt. Comm. J. Qual. Improv.*, Vol. 28, No. 10, 2002, pp. 531–545.
18. Mills, P. D., Neily, J., Luan, D., et al., Using Aggregate Root Cause Analysis to Reduce Falls and Related Injuries, *Jt. Comm. J. Qual. and Patient Saf.*, Vol. 31, No. 1, 2005, pp. 21–31.
19. Wald, H., Shojania, K. G., Root Cause Analysis, in *Making Health Care Safer: A Critical Analysis of Patient Safety Practices*, edited by A.J. Markowitz, Report No. 43, Agency for Health Care Research and Quality, U.S. Department of Health and Human Services, Rockville, Maryland, 2001, Chapter 5, pp. 1–7.
20. *Risk Analysis Requirements and Guidelines*, Report No. CAN/CSA-Q634-91, prepared by the Canadian Standards Association, 1991. Available from the Canadian Standards Association, 178 Rexdale Boulevard, Rexdale, Ontario, Canada.
21. Goetsch, D. L., *Occupational Safety and Health*, Prentice Hall, Englewood Cliffs, New Jersey, 1996.
22. Hammer, W., *Product Safety Management and Engineering*, Prentice Hall, Englewood Cliffs, New Jersey, 1980.
23. Dhillon, B. S., *Engineering Safety: Fundamentals, Techniques, and Applications*, World Scientific Publishing, River Edge, New Jersey, 2003.
24. Ericson, C. A., *Hazard Analysis Techniques for System Safety*, John Wiley and Sons, New York, 2005.
25. Vincoli, J. W., *Basic Guide to System Safety*, John Wiley and Sons, New York, 2006.
26. Goetsch, D. L., *Occupational Safety and Health*, Prentice Hall, Englewood Cliffs, New Jersey, 1996.
27. Hallock, R. G., Technic of Operations Review Analysis: Determine Cause of Accident/Incident, *Safety and Health*, Vol. 60, No. 8, 1991, pp. 38–39, 46.

28. Hammer, W., Price, D., *Occupational Safety Management and Engineering*, Prentice Hall, Upper Saddle River, New Jersey, 2001.
29. Tennant, G., *Six Sigma: SPC and TQM in Manufacturing and Services*, Gower, London, 2001.
30. Pande, P. S., Newman, R. P., Cavanaugh, R. R., *The Six Sigma Way*, McGraw-Hill, New York, 2000.
31. Hughes, R. G., Tools and Strategies for Quality Improvement and Patient Safety, in *Handbook for Nurses*, edited by R.G. Hughes, Agency for Healthcare Research and Quality, Rockville, Maryland, Chapter 44, pp. 1–22.
32. Barry, R., Murcko, A. C., Brubaker, C. E., *The Six Sigma Book for Healthcare: Improving Outcomes by Reducing Errors*, Health Administration Press, Chicago, 2003.
33. Pande, P. S., Newman, R. P., Cavanaugh, R. R., *The Six Sigma Way: Team Field Book*, McGraw-Hill, New York, 2002.
34. Lanham, B., Maxson-Cooper, P., Is Six Sigma the Answer for Nursing to Reduce Medical Errors and Enhance Patient Safety?, *Nursing Economics*, Vol. 21, No. 1, 2003, pp. 39–41.
35. Smith, G. M., *Statistical Process Control and Quality Improvement*, Prentice Hall, Upper Saddle River, New Jersey, 2001.
36. Dhillon, B. S., *Reliability, Quality, and Safety for Engineers*, CRC Press, Boca Raton, Florida, 2005.
37. Besterfield, D. H., *Quality Control*, Prentice Hall, Upper Saddle River, New Jersey, 2001.
38. Dhillon, B. S., *Reliability and Quality Control: Bibliography on General and Specialized Areas*, Beta Publishers, Gloucester, Ontario, Canada, 1992.
39. Mears, P., *Quality Improvement Tools and Techniques*, McGraw-Hill, New York, 1995.

5

Patient Safety Basics

5.1 Introduction

Patient safety is a relatively new health care discipline that clearly emphasizes the reporting, analysis, and prevention of all types of medical errors that frequently result in adverse health care events. The Institute of Medicine in the United States defines patient safety as "the prevention of harm to patients" [1,2].

The occurrence, frequency, and magnitude of avoidable adverse patient-related events were not well known until the 1990s, when a number of countries reported staggering numbers of patient harm and deaths due to human errors. For example, the United States lost more American lives to patient safety-associated adverse events every 6 months than it did during the entire Vietnam War [3].

Although many health care settings and hospitals already have well-developed procedures in place to ensure patient safety, the health care sector still lags considerably behind the industrial sectors that have introduced various types of systematic safety-related processes. Current patient safety initiatives include items such as applying lessons learned from business and industry, adopting modern innovative technologies, educating providers and the public, and improving medical error reporting systems. This chapter presents various basic aspects of patient safety.

5.2 Patient Safety Goals

Although patient safety goals can vary from one organization to another, in 2001 the Joint Commission on Accreditation of Healthcare Organizations (JCAHO) in the United States developed a number of patient safety goals with the aim of revising them annually as the need arises [4,5]. Currently, these goals cover nine distinct areas [6]: ambulatory health care, home care, critical access hospital, long-term care, office-based surgery, hospital, Medicare/Medicaid long-term care, laboratory, and behavioral health care. Some of the main JCAHO goals are shown in Figure 5.1 [4,7].

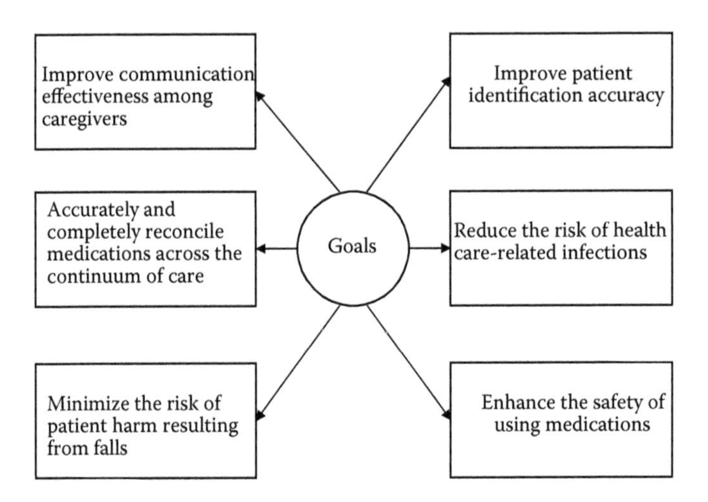

FIGURE 5.1
Some of the main JCAHO goals.

The goal "Improve communication effectiveness among caregivers" calls for actions such as the following:

- Measure, assess, and take necessary actions to enhance reporting timeliness as well as the timeliness of receipt by the responsible licensed caregiver of critical test results and values.
- Aim to standardize dose designations, acronyms, symbols, and abbreviations that are not to be used, within the organizational setup.
- In the case of telephone or verbal orders or for telephonic reporting of all types of critical test results, verify with care the entire order or test results by having the person receiving the transmitted information accurately record and "read back" the total order or test results in question.

The goal "Improve patient identification accuracy" is concerned with measures such as the use of a minimum of two patient identifiers to provide treatment, care, or services, and conducting a final verification process for confirming the right patient, procedure, and site prior to the initiation of any invasive procedure. The goal "Enhance the safety of using medications" is concerned with measures such as standardizing and limiting the number of drug concentrations used by the organization, labeling all medications and medication containers, and identifying and at least once a year reviewing a list of lookalike/soundalike drugs used by the organizational setup and taking necessary corrective actions for preventing errors involving the interchange of such drugs.

The goal "Reduce the risk of health care-related infections" is concerned with complying with the ongoing Centers for Disease Control and Prevention hand hygiene-related guidelines. The goal "Accurately and completely reconcile medications across the continuum of care" is concerned with actions such as those listed below:

- Develop a process to compare the patient's current medications with the ones ordered for the patient while under the care of the organization.
- Communicate the entire list of the patient's ongoing medications to the next service provider whenever a patient is referred or transferred to another practitioner, service, setting, or level of care within or outside the framework of the current organization.

Finally, the goal "Minimize the risk of patient harm resulting from falls" is basically concerned with the implementation of a fall reduction program that includes an assessment of the program effectiveness.

5.3 Causes of Patient Injuries and Examples of Factors Endangering Patient Safety at Various Levels of Health Care

Professionals working in the area of health care have identified many causes of patient injuries. Eight fundamental mechanisms (i.e., basic causes) through which a patient may be injured or killed are as follows [8]:

- Overdose
- Electrocution
- Suffocation/barotrauma
- Skin lesions (i.e., burns)
- Crushing
- Embolism
- Fire
- Performance failure

Additional information on these causes is available in Ref. [9].

There are many factors at various levels of health care endangering patient safety, some examples of which are presented below [9]:

- **Organization level.** Lack of leadership support to embrace a culture of safety.

- **Department level.** Lack of allocation of human and/or financial resources for supporting safety improvement projects.
- **Hospital level.** Lack of infrastructure for sustaining patient safety initiatives.
- **Team level.** Lack of communication or teamwork, lack of respect for each other's contribution to patient care, and misaligned goals.
- **Provider level.** Lack of receptiveness to new ideas, personal problems affecting concentration, and unwillingness to be a team player.
- **Patient level.** Failure to inform health care team of known allergies, reluctance to question unfamiliar actions, and patient-family conflict.
- **Work environment level.** Lack of access, lack of proper supplies, and insufficient preventive maintenance to ensure functioning equipment and devices.
- **Task level.** The complexity of care process itself (i.e., too many steps, high degree of difficulty, and interrelationships with other processes affecting the ability to complete task as intended).

5.4 Patient Safety Culture, Factors Contributing to Patient Safety Culture, and Its Assessment Objectives and Barriers

The history of the first serious attention paid to safety culture may be traced back to a report prepared by the International Nuclear Safety Advisory Group in 1988, on the Chernobyl nuclear power station disaster in Ukraine [10]. The concept has gained worldwide recognition in several industrial sectors, particularly in nuclear power generation and aviation. Probably the most important feature of the safety culture is shared perceptions among staff members and management concerning the importance of safety [11].

In regard to health care, safety culture has even greater importance, as safety applies not only to the workforce but also to the patients who may get injured due to staff actions. Various studies conducted over the years clearly indicate that the existence of a positive safety culture is essential for reducing preventable patient injuries and their cost to society at large [12–14].

Many factors have been highlighted as supporting the development of an effective patient safety culture. Some of the important ones are management (i.e., management commitment, ability, leadership, coordination, and flexibility), immediate supervisors (i.e., open-door policy, participation, and support correct behavior), reporting system (i.e., reporting near-miss, no-blame culture, analysis of error, open-door policy, confidentiality, and feedback), individual and behavioral (i.e., training, attitude, involvement, and behavior), and rules and procedures (i.e., clear and practical) [10,11,15–17].

There are many objectives that can be served by assessing an organization's safety culture, as listed here [13,18]:

- **Benchmarking.** This objective can be useful to evaluate the standing of a given unit in regard to a reference sample (i.e., comparable organizations, departments, and groups).
- **Measuring change.** This objective can be useful in applying and repeating over a period of time to detect changes in perceptions and attitudes.
- **Awareness enhancement.** This objective can be useful to increase awareness of staff members, particularly when performed in conjunction with other staff-oriented patient safety initiatives.
- **Accreditation.** This objective can be useful to serve as an element of a possibly mandated safety management review/accreditation program.
- **Profiling (diagnosis).** This objective can be useful in determining specific safety culture or climate profile of a unit, including the identification of "strong" and "weak" points.

A number of barriers to establishing patient safety culture have been identified by various organizations and professionals. The two main, probably the most important, ones are as follows [19]:

- Lack of awareness of the prevalence of adverse events in health care
- Denial of the severity of the problem

Additional information on these two and other barriers is available in Ref. [19].

5.5 Safer Practices for Better Health Care

In 2003 the National Quality Forum in the United States endorsed a total of 30 patient safety-related safer practices that should be implemented throughout clinical care settings to lower the risk of error and resultant harm to patients in general [20,21]. The first safer practice (i.e., Create and sustain a health care culture of safety) was composed of the following four basic components [20,21]:

- **Component 1.** Calls for health care organizations to measure their culture, provide appropriate feedback to the leadership and staff members, and undertake appropriate interventions that will decrease risk of patient safety.

- **Component 2.** Calls for health care organizations to systematically identify and mitigate patient safety-related hazards and risks with an effective integrated mechanism, in order to continuously reduce preventable patient-related harms.
- **Component 3.** Calls for health care organizations to develop an effective organization-wide program, in order to establish team-based care through teamwork training, skill building, and team-led performance improvement interventions that significantly decrease preventable harms to patients.
- **Component 4.** Calls for health care organizations to develop leadership structures and systems, for ensuring that there is organization-wide awareness of patient safety-related performance gaps and the accountability of involved leaders for the gaps.

Additional information on the remaining 29 patient safety-related safer practices endorsed by the National Quality Forum is available in Ref. [20].

5.6 Areas of Improved Patient Safety from Enhanced Health Information Exchange

Health care is an information-rich environment that requires many types of information even to make simple health care-related decisions. One of the most promising benefits of health-related information exchange is better patient safety. This means that enhanced health information exchange will directly improve patient safety because it will help to provide a better clinical picture of a given patient.

There are many areas in which patient safety can be improved from enhanced health information exchange, as shown in Figure 5.2 [22]. In the case of *improved medication information processing*, there are five main subareas for patient safety improvement from enhanced health information exchange. These subareas are drug–allergy information processing, drug–dose information processing, drug–drug information processing, drug–diagnosis information processing, and drug–gene information processing. Additional information on these five subareas is available in Ref. [22].

Improved laboratory information processing enabled by health information exchange can improve patient safety. Two basic areas for this are helping to ensure that lab testing under consideration is ordered and helping to ensure that lab test results are properly followed up on.

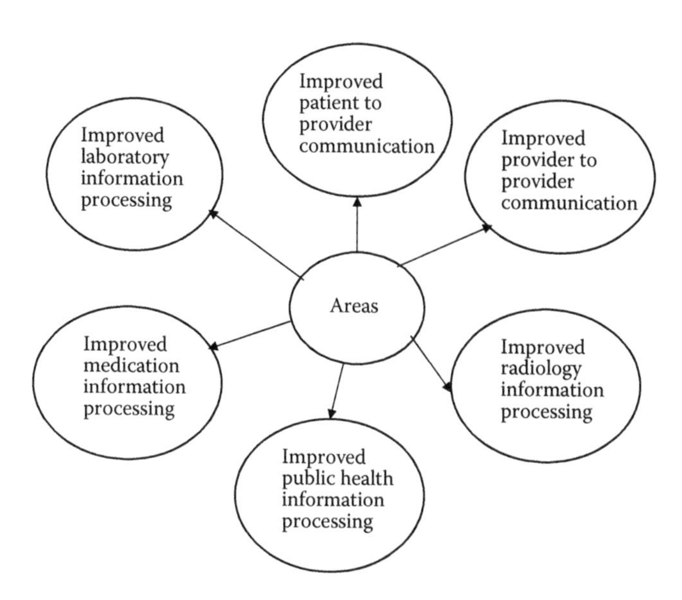

FIGURE 5.2
A number of areas of improved patient safety from enhanced health information exchange.

In the case of *improved patient-to-provider communication*, there are many ways in which health information exchange can improve patient safety, for example patients checking for errors in their medical history, reviewing medications and other health care instructions, being able to communicate more quickly with health care providers whenever they feel their safety may be at risk, and adding additional valuable information into their medical records. In the case of *improved provider-to-provider communication*, there are many health care-related scenarios in which patient safety is compromised because of poor health information exchange among providers. Providers who do not know patients, either in an outpatient or an inpatient setting, can make much safer decisions with improved health information exchange [22–24].

In the case of *improved radiology information processing*, generally the provider ordering an imaging study is different from the provider interpreting the imaging study, and patient safety can be improved in both these areas through enhanced health information exchange. For example, improved health information exchange can help to reduce adverse intravenous contrast reactions as well as reduce exposure to unnecessary radiology testing and radiation exposure. Finally, in the case of *improved public health information processing*, public health informatics is a rapidly growing area of health information exchange, in which opportunities for enhanced patient safety include postmarketing drug surveillance, environmental exposure surveillance, biohazard surveillance, and infections disease surveillance [22].

Additional information on the areas of improved patient safety from enhanced health information exchange is available in Ref. [22].

5.7 Patient Safety Program

The existence of a proper patient safety program is essential to guide the effective planning and implementation of safety-related projects. An eight-step patient safety program is presented below [7,25]:

- **Step 1:** Obtaining the feel for the beliefs or norms of the organization, department, unit, or team members. The survey responses can identify perceptions of staff members with respect to how important in their opinion safety is to the unit and to the organization.
- **Step 2:** Educating all members of the staff regarding the science of safety, so they understand better the reason that change is essential and the importance of their active participation in the program.
- **Step 3:** Determining the daily patient safety problems faced by the staff members. The problems for action are chosen on the basis of various factors including the frequency of occurrence, the potential for harm, the chances of developing and implementing a successful intervention, and the type of resources required for making the necessary change.
- **Step 4:** Forming an interdisciplinary team, once a safety initiative is identified, for taking a close look at the existing system and processes of care.
- **Step 5:** Planning and implementation of appropriate strategies for anticipating and preventing error occurrence, or for minimizing the potential for harm. The action plan incorporates within its framework a shared vision of appropriate goals that are focused, measurable, and simple.
- **Step 6:** Documenting the end results by following a systematic approach.
- **Step 7:** Sharing stories with others about the progress and the obstacles faced during the process. The presentation of a balanced picture of the effort can be a useful tool to spread improvements to other concerned areas and to keep the problem on the radar screen for the unit or organization.
- **Step 8:** Repeating Step 1 (i.e., conduct safety-related climate survey) to determine if there has been any shift in unit or organization culture toward one that values safety as its top priority.

5.8 Patient Safety Indicators

Patient safety indicators are a set of quality measures that makes use of hospital inpatient data to provide a perspective on patient safety. The indicators highlight difficulties that patients face through contact with the health care system and that are likely amenable to prevention by implementing appropriate changes at system level. The difficulties or problems highlighted are commonly known as adverse events or complications.

Some of the patient safety indicators are presented below [26,27].

5.8.1 Patient Safety Indicator I

This patient safety indicator is concerned with anesthesia reactions and complications and is expressed by

$$ARCI = \frac{\alpha_1}{\beta_1} \tag{5.1}$$

where

$ARCI$ = the anesthesia reactions and complications indicator.

α_1 = the number of adverse effects of or poisoning by anesthetic, endotracheal tube wrongly placed.

β_1 = the number of all surgical discharges.

Poisoning due to drug dependence or abuse and self-inflicting injury are not included in this safety indicator.

5.8.2 Patient Safety Indicator II

This patient safety indicator is concerned with foreign body left during procedure, and is expressed by

$$FBPI = \frac{\alpha_2}{\beta_2} \tag{5.2}$$

where

$FBPI$ = the foreign body left during procedure indicator.

α_2 = the number of times a foreign body was accidentally left during procedure.

β_2 = the number of all medical and surgical discharges.

5.8.3 Patient Safety Indicator III

This patient safety indicator is concerned with decubitus ulcer and is expressed by

$$DUI = \frac{\alpha_3}{\beta_3} \tag{5.3}$$

where

DUI = the decubitus ulcer indicator.
α_3 = the number of pressure ulcers.
β_3 = the number of all medical and surgical discharges with greater than 4 days' stay.

Hemiplegia, paraplegia or quadriplegia, paralysis, obstetric discharge, and admission from long-term care facilities are not included in this safety indicator.

5.8.4 Patient Safety Indicator IV

This patient safety indicator is concerned with infection due to medical care and is expressed by

$$IDMI = \frac{\alpha_4}{\beta_4} \tag{5.4}$$

where

$IDMI$ = the infection due to medical care indicator.
α_4 = the number of infections after infusion, injection, or transfusion, or as a result of vascular device or graft.
β_4 = the number of all medical and surgical discharges.

Cancer and immune compromise are not included in this safety indicator.

5.8.5 Patient Safety Indicator V

This patient safety indicator is concerned with transfusion reaction and is expressed by

$$TRI = \frac{\alpha_5}{\beta_5} \tag{5.5}$$

where

TRI = the transfusion reaction indicator.
α_5 = the number of secondary diagnoses of transfusion reaction.
β_5 = the number of all medical and surgical discharges.

5.8.6 Patient Safety Indicator VI

This patient safety indicator is concerned with postoperative respiratory failure and is expressed by

$$PRFI = \frac{\alpha_6}{\beta_6} \tag{5.6}$$

where
PRFI = the postoperative respiratory failure indicator.
α_6 = the number of postoperative acute respiratory failures.
β_6 = the number of all elective surgical discharges.

Respiratory or circulatory diseases and obstetric discharge are not included in this safety indicator.

5.8.7 Patient Safety Indicator VII

This patient safety indicator is concerned with technical difficulty with procedure and is expressed by

$$TDPI = \frac{\alpha_7}{\beta_7} \tag{5.7}$$

where
TDPI = the technical difficulty with procedure indicator.
α_7 = the number of accidental punctures or lacerations during procedure.
β_7 = the number of all medical and surgical discharges.

Obstetric discharges are not included in this safety indicator.

5.8.8 Patient Safety Indicator VIII

This patient safety indicator is concerned with deaths in low-mortality diagnosis-related group (DRG) and is expressed by

$$DLDI = \frac{\alpha_8}{\beta_8} \tag{5.8}$$

where

$DLDI$ = the deaths in low-mortality DRG indicator.

α_8 = the number of discharges with disposition of "deceased."

β_8 = the number of all discharges in DRGs with less than 5% mortality rate based on National Inpatient Sample data for 1997 (United States).

Trauma, immunocompromised state, and cancer are not included in this safety indicator.

5.8.9 Patient Safety Indicator IX

This patient safety indicator is concerned with postoperative hemorrhage/hematoma and is expressed by

$$PHI = \frac{\alpha_9}{\beta_9} \tag{5.9}$$

where

PHI = the postoperative hemorrhage/hematoma indicator.

α_9 = the number of postoperative hemorrhages/hematomas with surgical drainage or evacuation.

β_9 = the number of all surgical discharges.

Obstetric discharge is not included in this safety indicator.

5.8.10 Patient Safety Indicator X

This patient safety indicator is concerned with obstetric trauma (vaginal and instrumentation) and is expressed by

$$OTVI = \frac{\alpha_{10}}{\beta_{10}} \tag{5.10}$$

where

$OTVI$ = the obstetric trauma (vaginal with instrumentation) indicator.

α_{10} = the number of principal or secondary diagnoses of fourth-degree perineal, high vaginal, or cervical lacerations or procedures to repair any of these lacerations.

β_{10} = the number of all vaginal deliveries with forceps or vacuum.

5.8.11 Patient Safety Indicator XI

This patient safety indicator is concerned with iatrogenic pneumothorax and is expressed by

$$LPI = \frac{\alpha_{11}}{\beta_{11}} \tag{5.11}$$

where
LPI = the iatrogenic pneumothorax indicator.
α_{11} = the number of iatrogenic pneumothoraces.
β_{11} = the number of all medical and surgical discharges.

Trauma, cardiothoracic surgery, lung or pleural biopsy, and obstetric discharge are not included in this safety indicator.

5.8.12 Patient Safety Indicator XII

This patient safety indicator is concerned with postoperative pulmonary embolism or deep vein thrombosis and is expressed by

$$PPEI = \frac{\alpha_{12}}{\beta_{12}} \tag{5.12}$$

where
$PPEI$ = the postoperative pulmonary embolism or deep vein thrombosis indicator.
α_{12} = the number of postoperative deep vein thrombosis pulmonary embolisms.
β_{12} = the number of all surgical discharges.

Obstetric discharge, principal diagnosis of deep vein thrombosis, and obstetric patients are not included in this safety indicator.

5.8.13 Patient Safety Indicator XIII

This patient safety indicator is concerned with postoperative hip fracture and is expressed by

$$PHFI = \frac{\alpha_{13}}{\beta_{13}} \tag{5.13}$$

where
$PHFI$ = the postoperative hip fracture indicator.
α_{13} = the number of postoperative in-hospital hip fractures.
β_{13} = the number of all surgical discharges.

Musculoskeletal diseases, seizure, syncope, stroke, coma, cardiac arrest, anoxic brain injury, poisoning, delirium, trauma, self-inflicted injury, and cancers metastatic to bone are not included in this safety indicator.

5.8.14 Patient Safety Indicator XIV

This patient safety indicator is concerned with obstetric trauma (vaginal without instrumentation) and is expressed by

$$OTVOI = \frac{\alpha_{14}}{\beta_{14}} \tag{5.14}$$

where

$OTVOI$ = the obstetric trauma (vaginal without instrumentation) indicator.

α_{14} = the number of principal or secondary diagnoses of fourth-degree perineal, high vaginal, or cervical lacerations or procedures to repair any of these lacerations.

β_{14} = the number of all vaginal deliveries without forceps or vacuums.

5.8.15 Patient Safety Indicator XV

This patient safety indicator is concerned with birth trauma (injury to neonate) and is expressed by

$$BTI = \frac{\alpha_{15}}{\beta_{15}} \tag{5.15}$$

where

BTI = the birth trauma (injury to neonate) indicator.

α_{15} = the number of intracranial hemorrhage, extraclavicular fracture, spinal injury, nerve injury (other than facial and brachial plexus), and other birth trauma.

β_{15} = the number of all live births.

Preterm infants (for intracranial hemorrhage) and osteogenesis imperfecta (for fracture) are not included in this safety indicator.

5.8.16 Patient Safety Indicator XVI

This patient safety indicator is concerned with postoperative sepsis and is expressed by

$$PSI = \frac{\alpha_{16}}{\beta_{16}} \tag{5.16}$$

where

PSI = the postoperative sepsis indicator.

α_{16} = the number of postoperative sepsis.

β_{16} = the number of all elective surgical discharges with greater than 3 days' stay.

Cancer, infection, immune compromise, and obstetric discharges are not included in this safety indicator.

5.9 Problems

1. What are the main goals of the Joint Commission on Accreditation of Healthcare Organizations (JCAHO)?
2. Write an essay on patient safety.
3. What are the main causes of patient injuries?
4. Discuss examples of factors endangering patient safety at various levels of health care.
5. Discuss the factors that contribute to patient safety culture.
6. How many patient safety-related safer practices were endorsed by the National Quality Forum in the United States in 2003? Discuss one of these practices.
7. Discuss areas of improved patient safety from enhanced health information exchange.
8. Discuss patient safety program.
9. What are patient safety indicators?
10. Define at least five patient safety indicators.

5.10 References

1. Aspden, P., Corrigan, J., Wolcott, J., et al., Editors, *Patient Safety: Achieving a New Standard for Care,* National Academies Press, Washington, D.C., 2004.
2. Mitchell, P. H., Defining Patient Safety and Quality Care, in *Patient Safety and Quality: An Evidence-Based Handbook for Nurses*, edited by R. G. Hughes, AHRQ Publication No. 08-0043, Agency for Healthcare Research and Quality (AHRQ), U.S. Department of Health and Human Services, Rockville, Maryland, 2008, pp. 1–5.

3. *Patient Safety in American Hospitals*, Report, Health Grades, Golden, Colorado, July 2004.
4. Poe, S. S., Using Performance Improvement to Support Patient Safety, in *Measuring Patient Safety*, edited by R. Newhouse, S. S. Poe, Jones and Bartlett, Boston, 2005, pp. 13–25.
5. *National Patient Safety Goals*, The Joint Commission on Accreditation of Healthcare Organizations (JCAHO), 1 Renaissance Blvd., Oakbrook, Terrace, Illinois, 2007. Also available online at http://www.jointcommission.org/patientsafety/Nationalpatientsafety goals/07_npsg_facts.htm.
6. *National Patient Safety Goals*, The Joint Commission on Accreditation of Healthcare Organizations (JCAHO), 1 Renaissance Blvd., Oakbrook, Terrace, Illinois, 2010. Also available online at http://www.jointcommission.org/patientsafety/nationalpatientsafety goals/.
7. Dhillon, B. S., *Reliability Technology, Human Error, and Quality in Health Care*, CRC Press, Boca Raton, Florida, 2008.
8. Brueley, M. E., Ergonomics and Error: Who Is Responsible?, *Proceedings of the First Symposium on Human Factors in Medical Devices*, 1989, pp. 6–10.
9. Poe, S. S., Patient Safety: Planting the Seed, *Journal of Nursing Care Quality*, July–September, 2005, pp. 198–202.
10. Feng, X., Bobay, K., Weiss, M., Patient Safety Culture in Nursing: A Dimensional Concept Analysis, *Journal of Advanced Nursing*, Vol. 63, No. 3, 2008, pp. 310–319.
11. Clarke, S., Perceptions of Organizational Safety: Implications for the Development of Safety Culture, *Journal of Organizational Behaviour*, Vol. 20, 1999, pp. 185–188.
12. Zhan, C., Miller, M. R., Excess Length of Stay, Charges, and Mortality Attributable to Medical Injuries during Hospitalization, *Journal of the American Medical Association*, Vol. 290, 2003, pp. 1868–1874.
13. Madsen, M. D., Andersen, H. B., Itoh, K., Assessing Safety Culture and Climate in Healthcare, in *Handbook of Human Factors and Ergonomics in Healthcare and Patient Safety*, edited by P. Carayon, Lawrence Erlbaum Associates, Mahwah, New Jersey, 2007, pp. 693–713.
14. Kohn, L., Corrigan, J. M., Donaldson, M. S., *To Err Is Human: Building a Safer Health System*, Institute of Medicine, National Academy of Medicine, National Academies Press, Washington, D.C., 1999.
15. Cooper, M. D., Towards a Model of Safety Culture, *Safety Science*, Vol. 36, 2000, pp. 111–136.
16. Glendon, A. I., Stanton, N. A., Perspectives on Safety Culture, *Safety Culture*, Vol. 34, 2000, pp. 193–214.
17. Cohen, M., Eustis, M. A., Gribbins, R. E., Changing the Culture of Patient Safety: Leadership's Role in Health Care Quality Improvement, *Joint Commission Journal on Quality and Safety*, Vol. 29, 2003, pp. 329–335.
18. Nieva, V. F., Sorra, J., Safety Culture Assessment: A Tool for Improving Patient Safety in Healthcare Organizations, *Quality and Safety in Healthcare*, Vol. 12, Supplement II, 2003, pp. 17–23.
19. Frush, K. S., Alton, M., Frush, D. P., Development and Implementation of Hospital-Based Patient Safety Program, *Pediatric Radiology*, Vol. 36, 2006, pp. 291–298.

20. *Safety Practices for Better Healthcare: A Consensus Report*, The National Quality Forum, 601 Thirteenth Street, NW, Suite 500 North, Washington, D.C., 2003.
21. Frush, K. S., Fundamentals of a Patient Safety Program, *Pediatric Radiology*, Vol. 38, Supplement 4, 2008, pp. S685–S689.
22. Kaelber, D. C., Bates, D. W., Health Information Exchange and Patient Safety, *Journal of Biomedical Informatics*, Vol. 40, 2007, pp. S40–S45.
23. Sutcliffe, K. M., Lewton, E., Rosenthal, M. M., Communication Failures: An Insidious Contributor to Medical Mishaps, *Academic Medicine*, Vol. 79, No. 2, 2004, pp. 186–194.
24. Petersen, L. A., Orav, E. J., Teich, J. M., O'Neil, A. C., Brennan, T. A., Using a Computerized Sign-Out Program to Improve Continuity of Inpatient Care and Prevent Adverse Events, *Joint Commission Journal on Quality Improvement*, Vol. 24, No. 2, 1998, pp. 77–87.
25. Dawson, P. B., Moving Forward: Planning a Safety Project, in *Measuring Patient Safety*, edited by R. P. Newhouse and S. S. Poe, Jones and Bartlett, Boston, 2005, pp. 27–38.
26. Miller, M. R., Zhan, C., Pediatric Patient Safety in Hospitals: A National Picture in 2000, *Pediatrics*, Vol. 113, 2004, pp. 1741–1746.
27. Zhan, C., Miller, M. R., Administrative Data Based Patient Safety Research: A Critical Review, *Quality & Safety in Health Care*, Vol. 12, 2003, pp. ii58–ii63.

6

Medication Safety and Errors

6.1 Introduction

Over the past half-century a wide variety of medicinal drugs have been produced, and their usage has resulted in an impressive reduction in many potentially catastrophic conditions. Although there are many advantages of modern drug therapy, almost all medications possess the potential for harm from adverse drug events [1].

Adverse drug events may be defined as some kind of injuries resulting from medical interventions related to drugs [2]. They make up a spectrum of patient-related injuries originating from both nonpreventable and preventable causes and include all drug-associated injuries that result from adverse drug reactions, drug–drug interactions, or medication errors [3].

A medication error is an example of a preventable cause of an adverse drug event and is defined as any preventable event that may result in incorrect medication use or patient harm while the medication is in the control of a consumer, a health care professional, or a patient [4,5]. Medication-related errors result in deaths or serious injuries infrequently, but a sizable number of people each year are affected due to the widespread use of various types of drugs both outside and within the hospital environments [6].

This chapter presents various important aspects of medication safety and errors.

6.2 Medication Safety in Emergency Departments

Medication safety in emergency departments is a challenging issue; health care workers must function within an environment where multitasking is inevitable and sudden and frequent interruptions are normal. Ref. [7] has identified many factors in emergency medicine that result in adverse drug events. Drug ordering and delivery in emergency medicine plays an important role as it can, directly or indirectly, lead to harmful adverse drug events.

There are five stages of drug ordering and delivery in which safety checks/measures are very important. These stages are as follows [1]:

- **Prescribing stage.** This is the process concerned with choosing the right dose of the right drug at the right time by the right route for the correct diagnostic or therapeutic indications.
- **Transcribing stage.** This is concerned with communications between the prescriber and the individual dispensing or administering the medication.
- **Dispensing stage.** This is the process concerned with providing the correct medication to the individual who will look after administering the drug to the right patient.
- **Administration stage.** This is concerned with the administration of a drug, which is basically the act of physically placing the drug into the patient's body.
- **Monitoring stage.** This is concerned with the monitoring of patients after medication administration (i.e., immediately after drug administration and following discharge or admission for a period of time).

Safety checks/measures for each of the above five stages are presented below.

6.2.1 Prescribing Stage Safety Checks/Measures

Some of the important prescribing stage checks/measures are as follows [1,7]:

- Establish a proper drug reference system for assisting physicians, nursing personnel, and pharmacists in the correct dosages, uses, and applications of drugs or medications.
- When caring for pediatric patients, ensure that correct weight in kilograms is recorded on the prescription and on the patient's chart.
- Aim to make liberal use of pharmacists' expertise during the medication prescribing process, particularly when drugs are unfamiliar.
- Aim to order necessary laboratory studies for identifying patient characteristics that may place certain patients at risk for adverse drug events, and remain vigilant for any laboratory errors.
- Use extra caution during the medication prescribing process for pregnant patients.

6.2.2 Transcribing Stage Safety Checks/Measures

Some of the important transcribing stage safety checks/measures are as follows [1,7]:

- Aim to avoid using acronyms and abbreviations, and write neatly and clearly.

- Take appropriate measures to avoid confusion in written orders (e.g., trailing zeros).
- Ensure that a legible prescription contains information such as the patient's full name; the medication strength; for pediatric patient, his or her weight in kilograms; a list of instructions for the patient, including the purpose of medication; and the number or amount of medication to be dispensed.
- Make use of check-backs on verbal orders to ensure proper understanding.
- Always provide name and telephone number of the prescriber on the medication prescription, so that the pharmacists or nurses can clarify any confusion.

6.2.3 Dispensing Stage Safety Checks/Measures

Some of the important dispensing stage safety checks/measures are as follows [1,7,8]:

- Confirm with care placement of decimal points with the prescribing emergency physician.
- Ensure that, in the case of a high-alert medication, a second person performs the required arithmetic independently to confirm its accuracy.
- Ensure that the patient is not allergic to the medication being dispensed.
- Ensure that pharmacists are not filling incomplete prescriptions without consulting the prescribing physician.
- Check the patient's wristband to confirm the right patient prior to dispensing any medication.
- Ensure that all recorded weights are in kilograms, and confirm their accuracy.
- Ensure that all types of narcotics dispensed from the emergency department are recorded properly and all drugs dispensed from the emergency department are recorded in the patient's medical record.

6.2.4 Administration Stage Safety Checks/Measures

Some of the important administration stage safety checks/measures are as follows [1,7,9]:

- Aim not to stock in the emergency department certain high-alert medications such as potassium chloride.
- Always consult all appropriate reference material whenever in doubt about how to safely administer the medication in question.

- Ensure that there are no barriers to reporting administration-related errors, such as burdensome paperwork and fear of punitive action.
- Ensure that the high-alert medications such as heparin, potassium chloride, and nitroprusside are premixed in the pharmacy.
- Make use of appropriate safety checks to guarantee accuracy with respect to arithmetic errors, decimal point displacement, accurate weights, and confirming patient's allergy history.
- Consult with hospital pharmacists on a liberal basis with respect to factors such as drug concentrations, proper route of administration, and compatibility of agents being coadministered.

6.2.5 Monitoring Stage Safety Checks/Measures

Some of the important monitoring stage safety checks/measures are as follows [1,7]:

- Ensure that the patient is always monitored for an appropriate length of time in the emergency department after administration of medications, for effectiveness or for signs of adverse drug event occurrence.
- Ensure that all necessary follow-up arrangements are made for ensuring that proper monitoring is carried out.
- Ensure that written follow-up instructions are provided to all patients who have received drugs having amnestic properties.
- Ensure that all emergency department patients are informed in simple and straightforward language of the importance of all necessary monitoring tests.
- Ensure that all patients are informed appropriately of potentially serious side effects that mandate return to the emergency department.
- Ensure that all patients are provided with simple and straightforward written instructions about their medication's objective, side effects, and proper mode of administration.

6.3 Medication Safety in Operating Rooms

Medication safety in operating rooms is an important issue because medication-related problems can lead to serious adverse events. A study of medication in operating rooms reported the shortcomings in the following areas [10]:

- Patient information
- Drug information
- Communication of drug orders and information
- Drug labeling, packaging, and nomenclature
- Drug standardization, storage, and distribution
- Environment and workflow
- Staff competency and education
- Patient education
- Quality processes and risk management

The shortcomings in these areas include incomplete and inconsistent medication history in patient charts, pharmaceutical care not provided routinely, large number of abbreviations used on preprinted forms and in medication communications (verbal and written), medication brands changed without the knowledge of surgical teams or technicians, hazardous chemical found in close proximity to products designated for patient use, top of anesthesia carts cluttered, medication "stashes" found in selected areas, inconsistent preoperative teaching of patients, and inconsistent system of double-checks [10].

The corresponding study recommendations were to consistently document and complete preoperative medication history for all patients; to provide enhanced pharmacist support; to eliminate use of dangerous abbreviations and dose expressions; to enhance communication mechanisms; to evaluate need for, and then clearly identify and segregate, hazardous products; to minimize advance preparation of syringes for later administration and segregate them from the immediate workspace; to investigate, evaluate, and educate staff members about the dangers associated with workaround practices; to provide enhanced education materials for preoperative patients; and to consistently employ independent double-checks for hospital-selected "high-alert" drugs [10].

6.4 Drug Labeling or Packaging-Related Problems

A study of the United States Pharmacopoeia voluntary Medication Errors Reporting Program data over a one-year period revealed that in approximately 30% of the fatalities labeling or packaging was clearly cited as a contributory factor to medication errors that resulted in fatalities [11]. The types of labeling or packaging-related problems encountered are shown in Figure 6.1 [11].

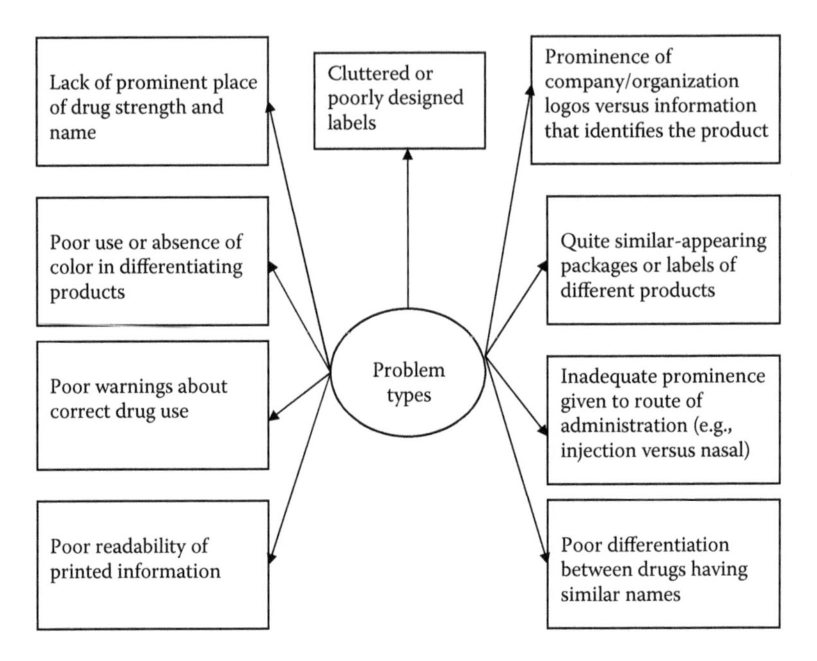

FIGURE 6.1
The types of labeling or packaging-related problems.

6.5 Prescribing Faults

Drug or medication prescribing faults may be grouped under five classifications, as shown in Figure 6.2 [12]. Underprescribing is concerned with failure to prescribe a drug or medication that is indicated and appropriate, or the use of too low a dose of an essential drug or medication. Some of the causes of underprescribing are as follows [12]:

- Fear of adverse effects or interactions
- Failure to recognize the appropriateness of therapy
- Doubts or ignorance about likely efficacy

Ineffective prescribing is concerned with prescribing a drug that is ineffective for the indication in general or for the patient in question. This is a pressing problem. For example, a study of 212 patients revealed that 6% of their 1621 medications were ineffective [13]. Overprescribing is concerned with prescribing a drug or medication in too high a dosage (i.e., too much, for too long, or too frequent).

Inappropriate prescribing and irrational prescribing are self-explanatory; additional information on them is available in Ref. [12].

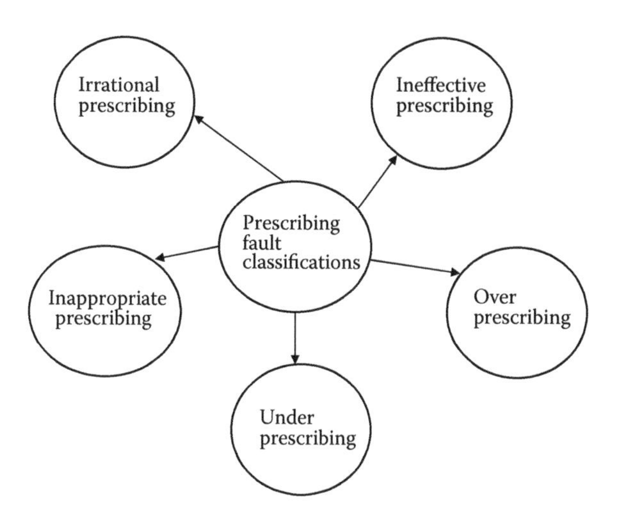

FIGURE 6.2
Prescribing fault classifications.

6.6 Medication-Use Safety Indicators

Over the years safety of medication use has become a serious issue. For example, each year approximately 1.5 million preventable adverse events occur in the United States, costing around $3.5 billion to its economy per annum [14]. Furthermore, according to a 2002 Commonwealth Fund survey, around 11% of patients in Canada clearly reported that they were given the incorrect medication at one time or another [14].

A number of indicators considered useful to improve medication-use safety, directly or indirectly, are presented below under six distinct categories [14].

6.6.1 Prescribing/Ordering Indicators

Five indicators belonging to this category are as follows.

6.6.1.1 Indicator I

This is expressed by

$$I_1 = \frac{\theta_1(100)}{\alpha_1} \tag{6.1}$$

where

I_1 = the value of indicator I expressed in percentage.

θ_1 = the number of prescriptions/medication orders using potentially dangerous dose abbreviations.
α_1 = the number of all prescriptions/medication orders.

6.6.1.2 Indicator II

This is expressed by

$$I_2 = \frac{\theta_2(100)}{\alpha_2} \tag{6.2}$$

where
I_2 = the value of indicator II expressed in percentage.
θ_2 = the number of prescriptions/medication orders using potentially dangerous medication abbreviations.
α_2 = the number of all prescriptions/medication orders.

6.6.1.3 Indicator III

This is expressed by

$$I_3 = \frac{\theta_3(100)}{\alpha_3} \tag{6.3}$$

where
I_3 = the value of indicator III expressed in percentage.
θ_3 = the number of prescriptions/medication orders with wrong leading and/or trailing zeros with decimal points.
α_3 = the number of all prescriptions/medication orders.

6.6.1.4 Indicator IV

This is expressed by

$$I_4 = \frac{\theta_4(100)}{\alpha_4} \tag{6.4}$$

where
I_4 = the value of indicator IV expressed in percentage.
θ_4 = the number of prescriptions/medication orders with "take as directed" as the only instruction for use.
α_4 = the number of all prescriptions/medication orders.

6.6.1.5 *Indicator V*

This is expressed by

$$I_5 = \frac{\theta_5(100)}{\alpha_5} \tag{6.5}$$

where

I_5 = the value of indicator V expressed in percentage.
θ_5 = the number of pediatric prescriptions for medications with a narrow therapeutic index with dose/weight calculations omitted.
α_5 = the number of all pediatric prescriptions for medications with a narrow therapeutic index.

6.6.2 Preparation and Dispensing Indicators

There is only one indicator that belongs to this category.

6.6.2.1 *Indicator I*

This is expressed by

$$I_6 = \frac{\theta_6(100)}{\alpha_6} \tag{6.6}$$

where

I_6 = the value of indicator I expressed in percentage.
θ_6 = the number of patient profiles in which allergy status is documented prior to dispensing the first prescription/medication order to the patient.
α_6 = the number of all patient profiles.

6.6.3 Administration Indicators

Three indicators belong to this category.

6.6.3.1 *Indicator I*

This is expressed by

$$I_7 = \frac{\theta_7(100)}{\alpha_7} \tag{6.7}$$

where

I_7 = the value of indicator I expressed in percentage.

θ_7 = the number of prescriptions/medication orders for high-alert medications that are double-checked and documented (with initials) by pharmacist prior to administration.

α_7 = the number of all prescriptions/medication orders for high-alert medications.

6.6.3.2 *Indicator II*

This is expressed by

$$I_8 = \frac{\theta_8(100)}{\alpha_8} \tag{6.8}$$

where

I_8 = the value of indicator II expressed in percentage.

θ_8 = the number of prescriptions/medication orders for high-alert medications using an administering protocol.

α_8 = the number of all prescriptions/medication orders for high-alert medications.

6.6.3.3 *Indicator III*

This is expressed by

$$I_9 = \frac{\theta_9(100)}{\alpha_9} \tag{6.9}$$

where

I_9 = the value of indicator III expressed in percentage.

θ_9 = the number of doses administered with machine-readable coding (i.e., bar codes).

α_9 = the number of all doses administered.

6.6.4 Monitoring/Assessment Indicators

Three indicators belong to this category.

6.6.4.1 *Indicator I*

This is expressed by

$$I_{10} = \frac{\theta_{10}(100)}{\alpha_{10}} \quad (6.10)$$

where

I_{10} = the value of indicator I expressed in percentage.
θ_{10} = the number of adverse drug event-related emergency room visits.
α_{10} = the number of all emergency room visits.

6.6.4.2 *Indicator II*

This is expressed by

$$I_{11} = \frac{\theta_{11}(100)}{\alpha_{11}} \quad (6.11)$$

where

I_{11} = the value of indicator II expressed in percentage.
θ_{11} = the number of adverse drug event–related hospitalizations.
α_{11} = the number of all hospitalizations.

6.6.4.3 *Indicator III*

This is expressed by

$$I_{12} = \frac{\alpha_{12}(100)}{\alpha_{12}} \quad (6.12)$$

where

I_{12} = the value of indicator III expressed in percentage.
θ_{12} = the number of beds with daily pharmacist participation in interdisciplinary direct patient care.
α_{12} = the number of all beds.

6.6.5 Purchasing/Inventory Management Indicators

There is only one indicator that belongs to this category.

6.6.5.1 *Indicator I*

This is expressed by

$$I_{13} = \frac{\theta_{13}(100)}{\alpha_{13}} \tag{6.13}$$

where

I_{13} = the value of indicator I expressed in percentage.
θ_{13} = the number of high-alert-prescription medications that are differentiated from other medications using flags, highlighting, or some other system.
α_{13} = the number of all high-alert prescription medications.

6.6.6 Systems of Care Indicators

Six indicators belong to this category.

6.6.6.1 Indicator I

This is expressed by

$$I_{14} = \frac{\theta_{14}(100)}{\alpha_{14}} \tag{6.14}$$

where

I_{14} = the value of indicator I expressed in percentage.
θ_{14} = the number of inpatients with complex high-risk medication regimens whose medication history was recorded on admission.
α_{14} = the number of all inpatients with complex high-risk medication regimens on admission.

6.6.6.2 Indicator II

This is expressed by

$$I_{15} = \frac{\theta_{15}(100)}{\alpha_{15}} \tag{6.15}$$

where

I_{15} = the value of indicator II expressed in percentage.
θ_{15} = the number of unintentional medication order discrepancies (e.g., omission, commission, wrong dose, and wrong frequency).
α_{15} = the number of all medication orders.

6.6.6.3 Indicator III

This is expressed by

$$I_{16} = \frac{\theta_{16}(100)}{\alpha_{16}} \tag{6.16}$$

where

I_{16} = the value of indicator III expressed in percentage.
θ_{16} = the number of patients whose medication profiles are reconciled within 24 hours of admission.
α_{16} = the number of all admitted patients.

6.6.6.4 Indicator IV

This is expressed by

$$I_{17} = \frac{\theta_{17}(100)}{\alpha_{17}} \tag{6.17}$$

where

I_{17} = the value of indicator IV expressed in percentage.
θ_{17} = the number of patients whose medication profiles are reconciled within 24 hours before hospital discharge.
α_{17} = the number of all discharged patients.

6.6.6.5 Indicator V

This is expressed by

$$I_{18} = \frac{\theta_{18}(100)}{\alpha_{18}} \tag{6.18}$$

where

I_{18} = the value of indicator V expressed in percentage.
θ_{18} = the number of discharge medication summaries sent to community physicians within 72 hours of hospital discharge.
α_{18} = the number of all discharged patients on medications.

6.6.6.6 Indicator VI

This is expressed by

$$I_{19} = \frac{\theta_{19}(100)}{\alpha_{19}} \tag{6.19}$$

where

I_{19} = the value of indicator VI expressed in percentage.
θ_{19} = the number of discharge medication summaries sent to a community pharmacy within 72 hours of hospital discharge.
α_{19} = the number of all discharged patients on medications.

6.7 Medication Error-Related Facts and Figures

Some of the medication error-related facts and figures are as follows:

- The number of deaths from medication errors and adverse reactions to various types of medicines used in the U.S. hospitals increased from 2876 to 7391 during the period 1983–1993 [12-15].
- The annual cost of medication-related errors in the United States is estimated to be over $7 billion [5,16].
- In 1993, medication errors caused 7391 deaths in the United States [5,6,15].
- The cost of hospital-based medication errors is approximately $2 billion per year in the United States [17].
- A study of 36,200 medication orders conducted in the United Kingdom (UK) revealed that 1.5% contained prescribing errors [12,18].
- The number of annual deaths from medication errors in the UK increased from about 20 in 1990 to just under 200 in 2000 [12].
- According to Refs. [6,15], in 1993 over a 10-year period inpatient deaths due to medication errors increased by 2.37-fold in the United States as opposed to a 8.48-fold increase in outpatient deaths.
- A study conducted in the United States revealed that around 1.7% of prescriptions dispensed from community pharmacies contained errors to a certain degree [19].
- Medication errors range from around 5.3% to 20.6% of all administered doses [20].

6.8 Medication Errors Types and Common Reasons for Their Occurrence

There are basically the following eight types of medication errors [4,5,21]:

- **Incorrect-dose error.** The administration to patients of a dose that is lower or greater than the legitimate prescribers' recommended amount.
- **Omission error.** The oversight in the administration of a recommended dose to a patient before the next scheduled dose (if any).
- **Incorrect-time error.** The administration of medication outside a predefined interval of time from its scheduled administration time.
- **Unauthorized-drug error.** The administration of medication not recommended by the legitimate prescriber of a patient.
- **Incorrect-drug-preparation error.** Incorrect formulation or manipulation of the drug product prior to its administration.
- **Incorrect-administration-method error.** Incorrect method or procedure followed in the administration of a drug.
- **Incorrect-dosage-form error.** The administration to a patient of varying dosage of a drug, other than the dosage recommended by the legitimate prescriber.
- **Prescribing error.** Incorrect drug selection, dose, rate of administration, concentration, quantity, route, dose form, or instructions for use of a drug product authorized by its legitimate prescriber. Prescribing error can also be an illegible prescription. Some typical examples of a prescribing error are shown in Figure 6.3 [22].

There are many reasons for the occurrence of medication errors. Some of the common ones are illegible handwriting, incorrect transcriptions, errors in labeling, wrong abbreviations used during the prescribing process, incorrect dosage calculation, equipment failure, designation of ambiguous strength on labels or in packaging, lapses in individual performance, drug product

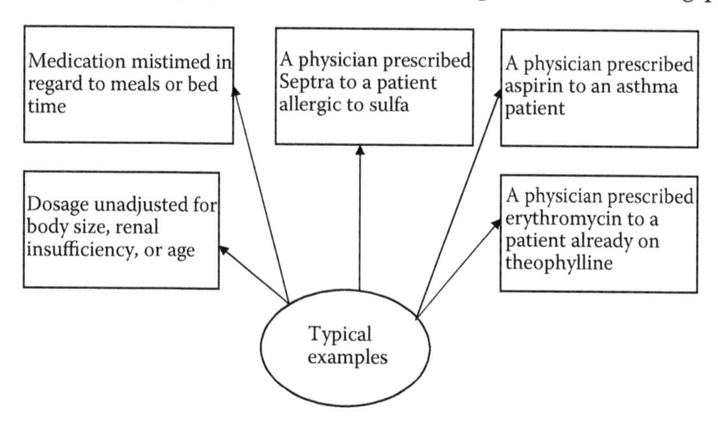

FIGURE 6.3
Some typical examples of a prescribing error.

nomenclature (e.g., use of numbered or lettered suffixes and prefixes in drug names), excessive workload, and poorly trained personnel [4].

6.9 Nursing-Related Factors in Medication Errors Occurrence

A significant proportion of medication errors can be attributed to the area of nursing. Some of the main factors that play an important role in the occurrence of nursing-related medication errors are shown in Figure 6.4 [23].

Nurses' knowledge of medication is an important factor because their lack of knowledge appears to be a continuing problem in the occurrence of various types of medication errors. For example, according to Ref. [24] approximately 25% of the 334 medication errors investigated were the result of the lack of proper drug knowledge among the nursing staff, pharmacists, and physicians.

Written prescription quality is another important factor, because often the nursing staff comes across poorly written and even illegible prescriptions that generate a potential for the occurrence of medication errors. Nurses' workload and staffing level is also an important factor because the workload carried by the nurses and the shortage of nursing manpower affect the rate of occurrence of medication errors [23–25]. The failure to adhere to policies

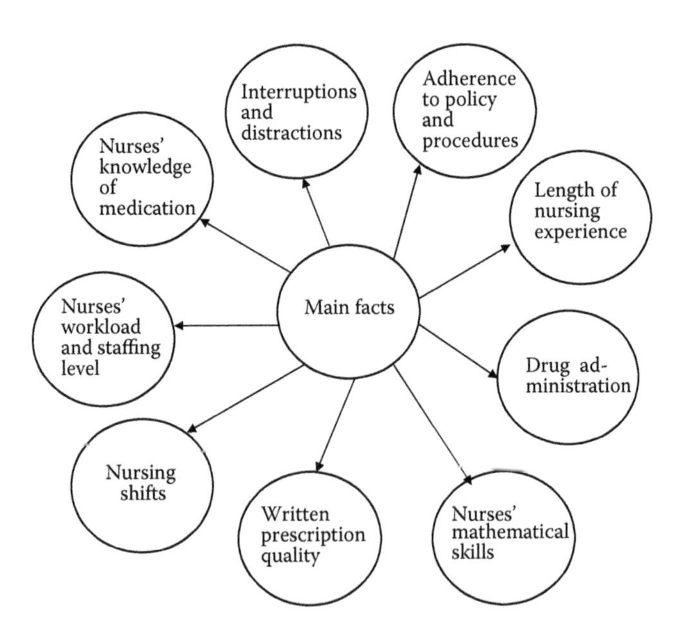

FIGURE 6.4
Some of the main nursing-related factors in the occurrence of medication errors.

and procedures in hospitals and other facilities by the nurses is a pressing problem in regard to medication errors [23,25]. For example, as per Ref. [26], approximately 72% of medication errors can be attributable to the failure of medical staff to follow proper policies and procedures.

Nursing shifts is an important factor as shown by various studies [23,27,28] conducted over the years that indicate that a variety of working conditions, including shift rotation, directly or indirectly contribute to the occurrence of various types of medication errors.

Nurses' mathematical skills are another important factor, as mathematical skills are essential to perform nursing functions such as intravenous regulations and medication calculations [29]. Poor mathematical skills of the nursing staff have resulted in various types of medication errors [29,30]. Often "distractions and interruptions" are cited as one of the main factors in the occurrence of nursing staff–related medication errors [23,31]. A survey of 175 nurses revealed that around 32% of them believe that frequent distractions or interruptions contribute to the occurrence of medication errors by the nursing staff [25].

Current research in regard to nursing experience and medication errors appears to be inconclusive, but studies indicate that nursing staff new to health care facilities are more likely to make errors, probably due to the new environment [5,23]. Finally, two-nurse drug administration generates fewer errors in comparison to single-nurse drug administration [32].

6.10 General Guidelines to Reduce the Occurrence of Medication Errors

Various guidelines have been developed over the years to reduce the occurrence of medication errors; some of the general ones are as follows [4,5,22]:

- Always check the identification bracelet of every patient before administering any medication.
- Provide concise and clear written and verbal medication-related instructions.
- Always perform medication dosage-related calculations on a piece of paper, not in the head.
- Make use of computer-generated prescriptions or write legibly.
- Always check the drug label three times (i.e., at the time of removing the container from storage, before administering the drug, and prior to returning or discarding the container).
- Pay close attention to both efficacy and safety when determining the amount of drug to be prescribed.

- Avoid all distractions during the preparation of medication for administration.
- Consider the possibility of inadvertent drug substitutions when the patient reports some side effects.
- Ask the patient about his/her allergies before administering any type of medication.
- Gain knowledge of the patient's diagnosis to ensure drug correctness.

6.11 Problems

1. Write an essay on medication safety and errors.
2. Discuss medication safety in emergency departments.
3. Describe at least four important drug prescribing stage safety checks/measures.
4. What are the medication-related shortcomings in operating rooms?
5. List at least nine types of drug labeling or packaging-related problems.
6. What are the five main classifications of medication prescribing–related faults?
7. Define two indexes for use in the area of prescribing/ordering.
8. List at least seven facts and figures concerned with medication errors.
9. What are the common reasons for the occurrence of medication errors?
10. List at least eight main nursing-related factors in the occurrence of medication errors.

6.12 References

1. Peth, H. A., Medication Safety in the Emergency Department, in *Patient Safety and Quality: An Evidence-Based Handbook for Nurses*, edited by R. G. Hughes, Agency for Healthcare Research and Quality, Rockville, Maryland, 2008, Chapter 21, pp. 1–10.
2. Bates, D. W., et al., Relationship between Medication Errors and Adverse Drug Events, *Journal of General Internal Medicine*, Vol. 10, 1995, pp. 199–205.
3. Hafner, J. W., et al., Adverse Drug Events in Emergency Department Patients, *Annals of Emergency Medicine*, Vol. 39, No. 3, 2002, pp. 258–267.

4. Coleman, J. C., Pharm, D., Medication Errors: Picking Up the Pieces, *Drug Topics*, March 1999, pp. 83–92.
5. Dhillon, B. S., *Human Reliability and Error in Medical System*, World Scientific Publishing, River Edge, New Jersey, 2003.
6. Kohn, L., Corrigan, J., Donaldson, M., Editors, *To Err Is Human: Building a Safer Health System*, National Academy Press, Washington, D.C., 1999.
7. Peth, H. A., Medication Errors in the Emergency Department: A Systems Approach to Minimizing Risk, in *High Risk Presentations in Emergency Medicine*, edited by H. A. Peth, W. B. Saunders, Philadelphia, 2003, pp. 141–158.
8. Poon, E. G., et al., Medication Dispensing Errors and Potential Adverse Drug Events Before and After Implementing Bar Code Technology in the Pharmacy, *Annals of Internal Medicine*, Vol. 145, 2006, pp. 426–434.
9. Kopp, B. J., et al., Medication Errors and Adverse Drug Events in an Intensive Care Unit: Direct Observation Approach for Detection, *Critical Care Medicine*, Vol. 34, No. 2, 2006, pp. 415–425.
10. Merali, R., et al., Medication Safety in the Operating Room: Teaming up to Improve Patient Safety, *Healthcare Quarterly*, Vol. 11, Special Issue, 2008, pp. 54–57.
11. Momtahan, K., et al., Using Human Factors Methods to Evaluate the Labelling of Injectable Drugs, *Healthcare Quarterly*, Vol. 11, Special Issue, 2008, pp. 122–128.
12. Aronson, J. K., Medication Errors: What They Are, How They Happen, and How to Avoid Them, *Quarterly Journal of Medicine*, Vol. 102, 2009, pp. 1093–1102.
13. Bregnhoj, L., et al., Prevalence of Inappropriate Prescribing in Primary Care, *Pharmacy World & Science*, Vol. 29, 2007, pp. 109–115.
14. Nigam, R., et al., Development of Canadian Safety Indicators for Medication Use, *Healthcare Quarterly*, Vol. 11, Special Issue, 2008, pp. 47–53.
15. Phillips, D. P., Christenfeld, N., Glynn, L. M., Increase in US Medication-Error Deaths between 1983–1993, *Lancet*, Vol. 351, 1998, pp. 643–644.
16. Wechsler, J., Manufacturers Challenged to Reduce Medication Errors, *Pharmaceutical Technology*, February 2000, pp. 12–22.
17. Smith, D. L., Medication Errors and DTC Ads, *Pharmaceutical Executive*, February 2000, pp. 129–130.
18. Dean, B., et al., Prescribing Errors in Hospital Inpatients: Their Incidence and Clinical Significance, *Quality & Safety in Health Care*, Vol. 11, 2002, pp. 340–344.
19. Flynn, E. A., Barker, K. N., Carnahan, B. J., National Observational Study of Prescription Dispensing Accuracy and Safety in 50 Pharmacies, *Journal of the American Pharmacists Association*, Vol. 43, 2003, pp. 191–200.
20. Bindler, R., Bayne, T., Medication Calculation Ability of Registered Nurses, *IMAGE*, Vol. 23, No. 4, 1991, pp. 221–224.
21. ASHP Guidelines on Preventing Medication Errors in Hospitals, *American Journal of Hospital Pharmacology*, Vol. 50, 1993, pp. 305–314.
22. Fox, G. N., Minimizing Prescribing Errors in Infants and Children, *American Family Physician*, Vol. 53, No. 4, 1996, pp. 1319–1325.
23. O'Shea, E., Factor Contributing to Medication Errors: A Literature Review, *Journal of Clinical Nursing*, Vol. 8, 1999, pp. 496–504.
24. Leape, L. L., et al., Systems Analysis of Adverse Drug Events, *JAMA*, Vol. 274, No. 1, 1995, pp. 35–43.
25. Conklin, D., et al., Medication Errors by Nurses: Contributing Factors, *AARN Newsletter*, Vol. 46, No. 1, 1990, pp. 8–9.

26. Long, G., Johnson, C., A Pilot Study for Reducing Medication Errors, *Quality Review Bulletin*, Vol. 7, No. 4, 1981, pp. 6–9.
27. Gold, D. R., et al., Rotating Shift Work, Sleep, and Accidents Related to Sleepiness in Hospital Nurses, *American Journal of Public Health*, Vol. 82, No. 7, 1992, pp. 1011–1014.
28. Girotti, M. J., Medication Administration Errors in an Adult Intensive Care Unit, *Heart and Lung*, Vol. 16, No. 4, 1987, pp. 449–453.
29. Bindler, R., Bayne, T., Do Baccalaureate Students Possess Basic Mathematics Proficiency, *Journal of Nursing Education*, Vol. 23, No. 5, 1984, pp. 192–197.
30. Bayne, T., Bindler, R., Medication Calculation Skills of Registered Nurses, *Journal of Continuing Education in Nursing*, Vol. 19, No. 6, 1988, pp. 258–262.
31. Davis, N. M., Concentrating on Interruptions, *American Journal of Nursing*, Vol. 94, No. 3, 1994, pp. 14–15.
32. Kruse, H., et al., Administering Restricted Medications in Hospital: The Implications and Costs of Using Two Nurses, *Australian Clinical Review*, Vol. 12, No. 2, 1992, pp. 77–83.

7

Health Care Workers' Role and Safety and Falls

7.1 Introduction

Health care workers are an important element of the health care system. They play a crucial role in the success or failure of health care systems around the globe. These professionals place themselves at risk daily of contracting various types of life-threatening infections from blood-borne pathogens including hepatitis B, hepatitis C, and HIV. Furthermore, various types of injuries from needles and other sharp medical devices, along with accidental exposure to blood and body fluids from sprays and splashes are some of the most serious hazards faced by health care workers throughout the world.

According to Ref. [1], in 1997 more than 650,000 injuries and illnesses were reported in the United States health care sector alone. Many of the injuries were musculoskeletal-related and resulted from patient handling, and as high as 30% of all nurses reported back pain that limits their ability to work effectively [1].

Falls occurring in health care facilities are one of the important issues in the health care system, as they can affect both health care workers and patients. For example, according to Ref. [2], the number of patient-related falls in U.S. hospitals alone may reach more than 1 million per year in the first quarter of this century.

This chapter presents various important aspects of health care workers' role and safety and falls.

7.2 Technology Commonly Used by the Nursing Profession

Health care workers or professionals such as nurses play an important role in the health care system by using various types of technology. Patient care technology has become quite complex and has transformed the way nursing care is conceptualized and delivered. Patient care technologies used by the nursing profession range from relatively simple devices such as syringes

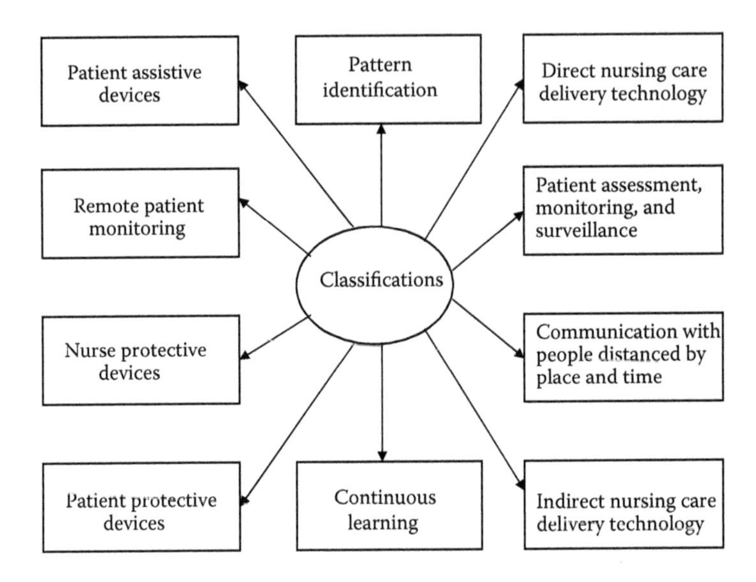

FIGURE 7.1
Classifications of technology commonly used by the nursing profession.

and catheters to rather complex systems such as bar code medication administration systems and electronic health records [3,4].

Technology commonly used by the nursing profession may be grouped under 10 distinct classifications, as shown in Figure 7.1 [4]. Direct nursing care delivery technology classification includes items such as needles, syringes, chest tubes, suction equipment, nebulizers, feeding pumps, traction systems, oxygen tanks and regulators, wound drainage tubes, oxygen and air regulators, oxygen tubing, face masks, intravenous (IV) tubing, endotracheal tubes, bar code medication administration, code carts, dressings (from gauze to specialized materials), ostomy appliances, urinary catheters and drainage bags, and tracheostomy tubes.

The patient assessment, monitoring, and surveillance classification includes items such as follows [4]:

- Thermometer
- Pulse oximetry
- Telemetry
- Ventilators
- Ostoscope
- Video surveillance
- Ophthalmoscope
- Stethoscope

- Bedside monitoring
- Sphygmomanometer

The patient assistive devices classification includes items such as walkers, bedpans, prosthetic limbs, canes, robotics, trapeze bars, wheelchair, orthotics (i.e., braces and shoes), patient transfer devices, and stand assist lifts. The six items belonging to the nurse protective devices classification are gloves, face masks, patient transfer devices, hand sanitizer dispensers, mechanical lifts, and gowns. The patient protective devices classification includes items such as listed below [4]:

- Hip protectors
- Fall alarms
- Limb compression devices
- Beds
- Specialized mattresses (e.g., low air loss)
- Specialized seating cushions
- Floor mats
- Hand rails in patient rooms, hallways, and bathrooms
- Specialized lighting
- Elopement/wandering alarms

The indirect nursing care delivery technology classification includes items such as robotics, computerized staffing systems, electronic inventory systems, and radio frequency identification. The three items belonging to the "continuous learning" classification are video conferencing, online learning, and distance learning. The communication with people distanced by place and time classification includes items such as follows [4]:

- Electronic ordering systems
- Communication devices (e.g., cell phones and paging systems)
- Electronic medical records
- Call systems, including emergency call bell

The two items belonging to the remote patient monitoring classification are telemedicine and telehealth. Finally, the pattern identification (to learn from errors and systems influences on adverse events) classification includes items such as electronic medical records and workload and staffing data systems. Additional information on technology commonly used by the nursing profession is available in Ref. [4].

7.3 Relationship between Nursing Workload and Patient Safety

The heavy workload of nurses (particularly in hospitals) is becoming a major problem for the U.S. health care system, as the demand for nurses is increasing because of population aging [5]. For example, during the period 2000–2020, the U.S. population over 65 years of age is expected to grow by 54% (i.e., 19 million) [6,7]. The four main reasons for nurses to have higher workloads than ever before are as follows [5]:

- Inadequate supply of trained nurses
- Increased demand for trained nurses
- Reduction in patient length of stay
- Reduction in staffing level and increased overtime

Six areas of relationship between nursing workload and patient safety that show how nursing workload can, directly or indirectly, affect patient safety are time, motivation, stress and burnout, errors in decision making (attention), violations or walk-around, and systemic/organizational impact [5]. In the case of time, the nursing staff with heavy workloads may not have adequate time to apply safe practices, conduct required tasks safely, or monitor patients, and may reduce their communication with concerned physicians and other providers—for example, no or little time for double-checking medications.

In the case of motivation, nursing staff members with heavy workloads may be dissatisfied with their job, thus affecting their motivation for high-quality performance level. The nursing staff with heavy workload may experience stress and burnout, which can negatively impact their performance—for example, reduced cognitive and physical resources available for nursing staff members to perform adequately. In the case of errors in decision making (attention), the high cognitive workload (one dimension of the nursing staff workload) of the nursing staff can contribute to the occurrence of errors such as slips and lapses or mistakes—for example, forgetting to administer medications to patients.

In the case of violations or work-around, heavy workload conditions may make it more difficult for nursing staff members to follow rules and guidelines properly, thus compromising the safety and quality of patient care; for example, inadequate hand washing. Finally, in the case of systemic/organizational impact, the heavy workload of a nursing supervisor or manager, a nurse, or another provider can affect the safety of patient care quality provided by another nurse. For example, a charge or supervising nurse may not be available to help other nursing staff members with their patients at a moment of need.

7.4 Health Care Worker Hazards

Each day health care workers are exposed to many types of hazards. The main ones are shown in Figure 7.2 [8]. These are infectious hazards, physical hazards, environmental hazards, chemical hazards, and psychosocial/psychological hazards (i.e., emotional stress). Infectious hazards are basically concerned with various types of infections. Health care workers are exposed routinely to many types of infectious hazards, particularly blood-borne pathogens. Over the past few decades, blood-borne pathogens have received increasing attention because of the focus on the human immunodeficiency virus (HIV), which can result in acquired immune deficiency syndrome (AIDS) [8].

Physical hazards are concerned with items such as back injury, electrocution, rapid pressure change, and violence. Environmental hazards include items such as ionizing radiation, nonionizing radiation, building-related illness, and noise.

Chemical hazards are concerned with items such as follows [8]:

- Aesthetic agents
- Antineoplastic agents
- Cleaning and sterilizing agents
- Medications

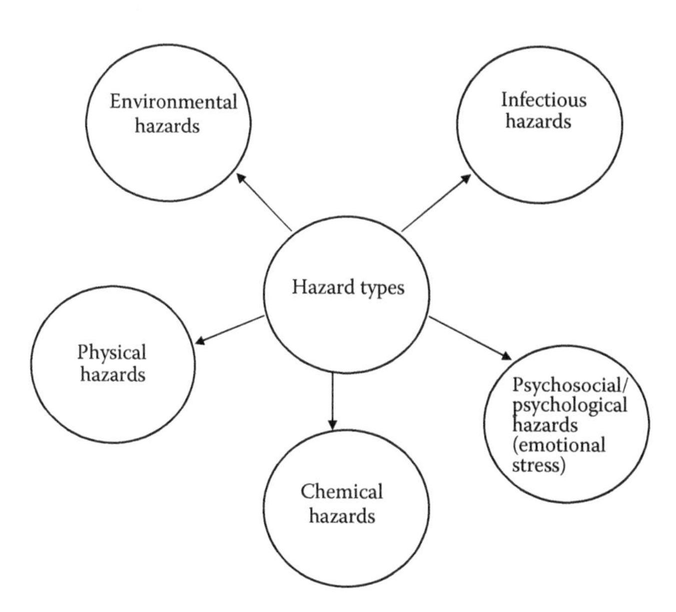

FIGURE 7.2
Main types of hazards of the health care profession.

- Insecticides
- Detergents
- Soaps

Finally, psychosocial/psychological hazards (i.e., emotion-related stresses) are considered one of the most pressing health hazards, with constant demands from the institution, patients, housekeepers, and coworkers such as nurses or doctors. Additional information on health care worker hazards is available in Refs. [8,9].

7.5 Health Care Worker Musculoskeletal and Needlestick Injuries and Latex Allergy

Musculoskeletal injuries are a leading category of injuries among health care workers. As per Ref. [1], out of more than 650,000 injuries and illnesses reported in the health care sector in 1997, most of the injuries were musculoskeletal-related, generally resulting from handling patients. In 2001, registered nurses in the United States had 108,000 work-related musculoskeletal injuries and disorders that involved lost work time, a rate similar to workers in the construction industry [10].

A recent study of 944 injuries reported by 23,742 health care workers that resulted in time-loss from work revealed that 83% of these injuries were musculoskeletal-related [11]. The occupational activities that resulted in musculoskeletal injuries were as follows [11]:

- Patient handling
- Patient care
- Material/equipment handling
- Equipment operation
- Office work
- Natural activity
- Other

Additional information on these activities is available in Ref. [11]. Some of the useful interventions for preventing musculoskeletal injuries associated with patient handling are using patient handling equipment/devices, training on proper use of patient handling equipment/devices, and following no-lift policies [12].

Needlestick injuries occur quite frequently among health care workers when they draw blood, administer an intravenous or intramuscular drug, or perform some other tasks that involve sharp instruments. This sets the stage for transmitting various types of viruses from the source individual to the recipient. Some facts directly or indirectly concerned with needlestick injuries are as follows:

- Each year around 800,000 needlestick injuries occur in the United States [1,12,13].
- Each year more than 1000 workers in the area of health care in the United States will contract a serious infection such as HIV or hepatitis B or C virus from needlestick injuries [12].
- In an average-size hospital in the United States, health care workers incur around 30 needlestick injuries per 100 beds per year [12].
- Nurses incur most needlestick injuries. Specifically, 54% of reported needlestick and sharp-object-related injuries involve the nursing staff [12,13].
- Around 50 to 247 health care workers get infected with hepatitis C virus (HCV) from work-related needlesticks each year in the United States [12].
- Each year needlestick injuries cause around 66,000 hepatitis B virus (HBV) infections, 16,000 HCV infections, and 1000 HIV infections worldwide [14].

Because needlestick injuries were a pressing problem, the U.S. Congress passed the Federal Needlestick Safety Act in 2000. This act not only requires employers to make safer needles available to health care workers but also requires employers to solicit the input of frontline health care workers when making needle purchasing-related decisions.

Some steps to prevent the occurrence of needlestick injuries are to reduce or eliminate the use of sharp instruments as much as possible, initiate appropriate engineering controls (i.e., needles with safety devices), initiate appropriate administrative controls including training and provision of adequate resources, and initiate appropriate work practice controls (e.g., avoiding hand-to-hand passing of sharp instruments and using instruments, not fingers, to grasp needles).

Latex allergy is a term that encompasses a range of allergic reactions to natural rubber latex and is a serious problem among health care workers. Health care workers who frequently use latex gloves and other latex-containing medical items are generally at risk of developing latex allergy. About 8–12% of health care workers react positively to blood test for latex allergy [1].

Some of the main factors that seem to indicate a higher risk for developing latex allergy are asthma, a history of allergies, and skin rashes. Individuals with multiple surgical procedures also appear to be at greater risk [1]. Actions suggested by the American Nurses Association to protect nurses and patients from latex allergy in all health care settings include the use of low-allergen powder-free gloves and removal of all latex-containing items from the worksite throughout the facility to reduce exposure [12].

Additional information on latex allergy is available in Refs. [1,12].

7.6 Health Care Workers' Slip-, Trip-, and Fall-Related Concerns and Prevention Strategies

The health care industry employs around 13 million workers in the United States and is probably the largest employer in the country [15]. It ranks second among eight industries as having the highest percentage of fall-related claim costs [15,16]. In 2002, over 296,000 injuries occurred in U.S. hospitals, more than in any other industrial sector in the country [17]. Furthermore, U.S. hospitals have a much greater than average rate of slips, trips, and falls on the same level [15].

As per 2007 Bureau of Labor statistics [15], the incidence rate of lost-work-day injuries from slips, trips, and falls on the same levels in U.S. hospitals was 35.2 per 10,000 full-time equivalents (FTEs)—about 75% greater than that of all other private industries combined (i.e., 20.2 per 10,000 FTEs).

Some of the main useful strategies for preventing health care worker slips, trips, and falls in the hospital environment are shown in Figure 7.3 [15]. Additional information on these strategies is available in Ref. [15].

7.7 Inpatient Bed Falls

Although less than 1% of inpatient falls in the United States result in fatalities, this still translates to around 11,000 annual fatal falls in the hospital environment in the country [2]. Bed falls are an important element of inpatient falls. As per Refs. [18–20] a number of studies in Western countries reported that about 42–60% of inpatient falls either were bed-related or involved patients who were discovered in the area around their bed after the occurrence of falls.

Some guidelines to prevent inpatient falls in bed area are as follows [21,22]:

- Nurse patients in beds that are a suitable length to patient height.
- Ensure frequently that all items used by patients are left within their reach.

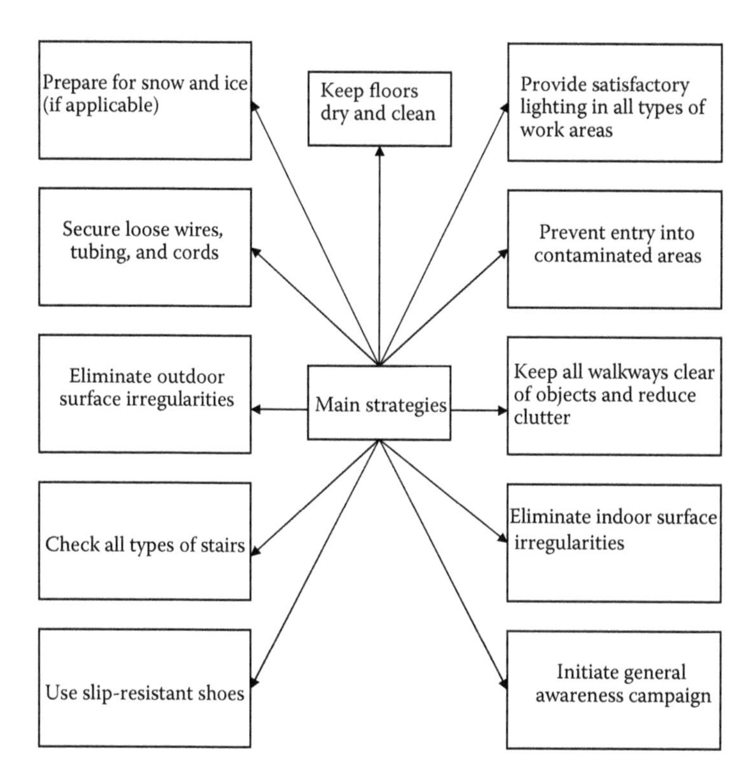

FIGURE 7.3
Some of the main strategies to prevent health care worker slips, trips, and falls in hospital environment.

- Ensure that when the patient bed height is raised for care, it is returned to a low position after completion of care.
- Ensure that the over-bed table is positioned on the non-exit side of the bed.
- Use appropriate night light for keeping the patient bed area illuminated at all times.
- Ensure that the patient bed brakes are on at all times.
- Place an antislip, absorbent mat on the floor (if available) in the area where the patient exits the bed.
- Place the commode chair (if in use) in the area next to the patient bed, on the exit side.
- Position all mobility-related aids (if appropriate) within easy reach of patients.
- Demonstrate the use of the call bell to all patients and ask them for a return demonstration. Place bell within reach when patient is in bed area.

7.8 Wheelchair-Related Falls

A large number of people use wheelchairs to move from one point to another in the United States. In fact, as per the 2000 census between 1.6 and 2.2 million people in the United States rely on wheelchairs to compensate for mobility impairments [23]. Adverse incidents associated with wheelchair use are not uncommon. As per Refs. [23,24], on the average over 36,500 wheelchair-related nonfatal accidents occur annually and approximately 1 wheelchair-related death occurs per week in the United States. The majority of these adverse events are directly or indirectly associated with falls.

The medical-related cost of wheelchair-related falls, including rehabilitation, can be between $25,000 and $75,000 [23]. Some of the key user characteristics associated with injuries from wheelchair-related accidents are as follows [25]:

- Daily use of a wheelchair
- Male gender and younger age
- Propelling with both hands
- Spina bifida or paraplegia as the reason for using wheelchair
- Wheelchair features: lightweight with adjustable axle, camber, and knapsack on back

There are many useful interventions to prevent the occurrence of wheelchair-related falls. They can be grouped under four categories, as shown in Figure 7.4 [23,25,26–30]. These categories are user precautions, environmental modifications, provider modifications, and wheelchair modifications.

User precautions includes items such as keep wheelchair in good condition, do not put heavy loads on the back of a manual wheelchair, avoid riding in the rain, avoid going down steep slopes, turn powered wheelchairs off prior to transferring, do not remove the anti-tippers, carefully read the operating manual and observe all precautions, avoid pulling backward on door or other objects when sitting in a manual wheelchair, and beware of caster flutter and have it fixed.

The *environmental modifications* include items such as widened doorways, easy-to-open doors, railings, and kitchen modifications. The *provider modifications* include items such as provide wheelchairs in specialized settings, provide appropriate wheelchair for the condition, heed safety-related warnings, and ensure satisfactory training.

Finally, *wheelchair modifications* include items such as automatic wheelchair locks, laptop safety cushions, seat belts, anti-tipping devices, transfer devices, frame modifications, and wheelchair monitors.

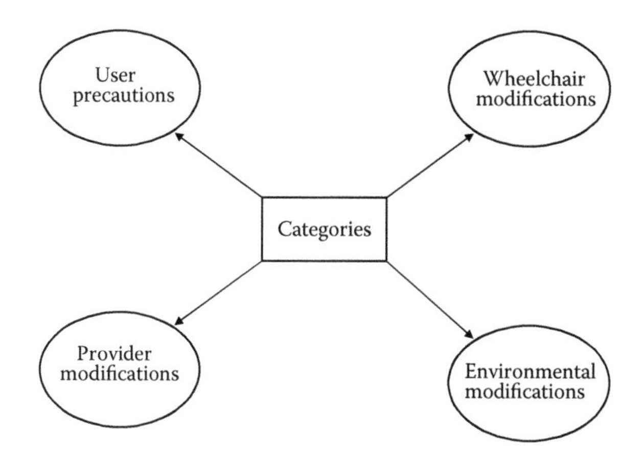

FIGURE 7.4
Categories of useful interventions for preventing wheelchair-related falls.

7.9 Fall Prevention Recommendations for Acute and Long-Term Care

Many fall prevention-related recommendations in the area of patient acute and long-term care have been proposed. Some of these recommendations are as follows [2]:

- Provide sufficient education to all staff members on safety care.
- Monitor with care all types of medication-related side effects and make adjustments as the need arises.
- Provide adequate training to all medical team members, including residents and students, on fall-injury risk assessment and on post-fall assessment.
- Provide appropriate exercise interventions to all long-term care patients.
- Adjust environments (e.g., design rooms for promoting safe patient movement).
- Treat underlying disorders such as anemia, diabetes, and syncope.
- Use appropriate alarm devices.

7.10 Problems

1. Write an essay on health care workers' role and safety.

2. Discuss the classifications of technology commonly used by the nursing profession.
3. What are the main reasons for nurses to have higher workloads than ever before?
4. Discuss the types of health care worker hazards.
5. List at least five facts and figures directly or indirectly concerned with the needlestick injuries of health care workers.
6. What is latex allergy?
7. What are the main strategies to prevent health care worker slips, trips, and falls in the hospital environment?
8. List at least 10 useful guidelines to prevent inpatient falls in bed area.
9. What are the main user characteristics associated with injurious wheelchair-related accidents?
10. List the elements of the following two categories of interventions for preventing wheelchair-related falls:
    - User precautions
    - Wheelchair modifications

7.11 References

1. Minter, S. G., Is Healthcare Safety Being Neglected?, *Occupational Hazards*, April 1999, pp. 37–42.
2. Currie, L., Fall and Injury Prevention, in *Patient Safety and Quality: An Evidence-Based Handbook for Nurses*, edited by R. G. Hughes, Agency for Healthcare Research and Quality, Rockville, Maryland, 2008, Chapter 10, pp. 1–27.
3. Hyman, W. A., Errors in Use of Medical Equipment, in *Human Error in Medicine*, edited by M. S. Bogner, Lawrence Erlbaum, Hillsdale, New Jersey, 1994, pp. 327–347.
4. Powell-Cope, G., Nelson, A. L., Patterson, E. S., Patient Care Technology and Safety, in *Patient Safety and Quality: An Evidence-Based Handbook for Nurses*, edited by R. G. Hughes, Agency for Healthcare Research and Quality, Rockville, Maryland, 2008, Chapter 50, pp. 1–13.
5. Carayon, P., Gurses, A. P., Nursing Workload and Patient Safety: A Human Factors Engineering Perspective, in *Patient Safety and Quality: An Evidence-Based Handbook for Nurses*, edited by R. G. Hughes, Agency for Healthcare Research and Quality, Rockville, Maryland, 2008, Chapter 30, pp. 1–10.
6. *Nursing Workforce: Emerging Nurse Shortages Due to Multiple Factors*. Report No. GAO-01-944, United States General Accounting Office (GAO), Washington, D.C., 2001.
7. *Nursing Workforce: Recruitment and Retention of Nurses and Nurse Aides Is a Growing Concern*, Report No. GAO-01-750T, United States General Accounting Office, Washington, D.C., 2001.

8. Behling, D., Guy, J., Industry Profile: Healthcare—Hazards of the Healthcare Profession, *Occupational Health and Safety*, Vol. 62, No. 2, 1993, pp. 54–57.
9. Clever, L. H., Omenn, G. S., Hazards for Health Care Workers, *Annual Review of Public Health*, Vol. 9, 1988, pp. 273–303.
10. *Lost-Work Time Injuries and Illnesses: Characteristics and Resulting Time Away from Work*, Report No. 04-460, Bureau of Labor and Statistics, U.S. Department of Labor, Washington, D.C., 2002.
11. Ngan, K., et al., Risks and Causes of Musculoskeletal Injuries among Healthcare Workers, *Occupational Medicine*, Vol. 60, 2010, pp. 389–394.
12. Trinkoff, A. M., et al., Personal Safety for Nurses, in *Patient Safety and Quality: An Evidence-Based Handbook for Nurses*, edited by R. G. Hughes, Agency for Healthcare Research and Quality, Rockville, Maryland, 2008, Chapter 39, pp. 1–23.
13. Henry, K., Campbell, S., Needle Stick/Sharps Injuries and HIV Exposure among Healthcare Workers: National Estimates Based on a Survey of U.S. Hospitals, *Minnesota Medicine*, Vol. 78, No. 11, 1995, pp. 41–44.
14. Pruss-Ustun, A., Rapiti, E., Hutin, Y., Estimation of the Global Burden of Disease Attributable to Contaminated Sharps Injuries among Healthcare Workers, *American Journal of Industrial Medicine*, Vol. 48, 2005, pp. 482–490.
15. Bell, J. L., et al., Evaluation of a Comprehensive Slip, Trip, and Fall Prevention Programme for Hospital Employees, *Ergonomics*, Vol. 51, No. 12, 2008, pp. 1906–1925.
16. Cotnam, J. P., Chang, W. R., Courtney, T. K., A Retrospective Study of Occupational Slips, Trips, and Falls across Industries, *Proceedings of the 44th Annual International Ergonomics Association/Human Factors Ergonomics Congress*, 2000, pp. 473–476.
17. Wiatrowski, W. J., Comparing Old and New Statistics on Workplace Injuries and Illnesses, *Monthly Labor Review*, December 2004, pp. 10–24.
18. Tzeng, H. M., Yin, C. Y., Heights of Occupied Beds: A Possible Risk Factor for Inpatient Falls, *Journal of Clinical Nursing*, Vol. 17, 2008, pp. 1503–1509.
19. Fonda, D., et al., Sustained Reduction in Serious Fall-Related Injuries in Older People in Hospital, *Medical Journal of Australia*, Vol. 184, 2006, pp. 379–382.
20. Masud, T., *Audit of Falls in the Medical Directorate*, Internal Report, Nottingham City Hospital, Nottingham, United Kingdom, 2003.
21. Clark, G. A., A Study of Falls among Elderly Hospitalized Patients, *Australian Journal of Advanced Nursing*, Vol. 2, 1985, pp. 34–44.
22. Hendrich, A., et al., Hospital Falls: Development of Predictive Model for Clinical Practice, *Applied Nursing Research*, Vol. 8, 1995, pp. 129–139.
23. Gavin-Dreschnack, D., et al., Wheelchair-Related Falls: Current Evidence and Directions for Improved Quality Care, *Journal of Nursing Care Quality*, Vol. 20, 2005, pp. 119–127.
24. Kirby, R. L., Ackroyd-Stolarz, S. A., Wheelchair Safety—Adverse Reports to the United States Food and Drug Administration, *American Journal of Physical Medicine & Rehabilitation*, Vol. 74, No. 4, 1995, pp. 308–312.
25. Kirby, R. L., et al., Wheelchair-Related Accidents Caused by Tips and Falls among Non-Institutionalized Users of Manually Propelled Wheelchairs in Nova Scotia, *American Journal of Physical Medicine & Rehabilitation*, Vol. 73, No. 5, 1994, pp. 319–330.
26. Joanna Briggs Institute for Evidenced Based Nursing, Falls in Hospitals, *Best Practices*, Vol. 2, No. 2, 1998, pp. 1–6.

27. Kirby, R. L., et al., Could Changes in the Wheelchair Delivery System Improve Safety?, *CMAJ*, Vol. 153, No. 11, 1995, pp. 1585–1591.
28. Knowlton, L., Preventing Falls, Improving Outcomes, *Geriatric Times*, Vol. 2, No. 5, 2001, pp. 1–6.
29. Brechtelsbauer, D. A., Louie, A., Wheelchair Use among Long-Term Care Residents, *Annals of Long Term Care*, Vol. 7, No. 6, 1999, pp. 213–220.
30. Berg, K., Hines, M., Allen, S., Wheelchair Users at Home: Few Home Modifications and Many Injurious Falls, *American Journal of Public Health*, Vol. 92, No. 1, 2002, pp. 48–49.

8

Human Error in Various Medical Areas and Other Related Information

8.1 Introduction

Human error is an important factor in many different medical areas that, directly or indirectly, influences patient safety. These areas include anesthesia, emergency medicine, operating rooms, intensive care units, medical technology use, laboratory testing, radiotherapy, image interpretation, and surgical pathology. For example, over 90% of the adverse events that occur in emergency departments are considered preventable [1–3], and a study of critical incident reports in an intensive care unit over a 10-year period (1989–1999) reported that most of the incidents were the result of staff errors [4].

Over the years the cost of human errors in the U.S. health care system has skyrocketed. The annual national cost of adverse events is about $38 billion, around $17 billion of which is associated with preventable adverse events [5]. Furthermore, the cost of adverse drug events alone in a U.S. hospital is approximately $5.6 million per year; about $2.8 million of this amount is associated with preventable adverse drug events [6].

This chapter presents various important aspects of human error in various medical areas and important related information.

8.2 Human Error in Anesthesia

The occurrence of human errors in anesthesia has become an important issue because in recent years many anesthesia-related deaths are considered to be the result of human error. A study of 589 anesthesia-related deaths revealed that human error was an important factor in 83% of the cases [7,8]. Additional facts and figures directly or indirectly concerned with human error in anesthesia are available in Ref. [2].

Some of the common anesthesia errors are as follows [9]:

- Drug overdose
- Loss of oxygen supply

- Syringe swap
- Inadvertent change in gas flow
- Breathing circuit misconnection
- Ventilator failure
- Ampule swap
- Breathing circuit disconnection
- Hypoventilation (operator error)
- Premature extubation
- Endobronchial intubation
- Breathing circuit leak
- Unintentional extubation
- Wrong selection of airway management method

Additionally, some of the common causes of anesthesia errors are carelessness, haste and fatigue, emergency case, poor familiarity with anesthetic method, inadequate anesthesia-related experience, lack of skilled assistance or supervision, excessive reliance on other personnel, failure to carry out a proper examination, visual field restriction, poor familiarity with equipment/device, teaching activity in progress, and poor communication with the surgical team or the laboratory personnel [10].

8.2.1 Methods for Reducing or Preventing Human Error-Related Anesthetic Mishaps

Many methods have been developed to reduce or prevent the occurrence of human error-related anesthetic mishaps. Two of these methods are presented below [2].

8.2.1.1 Method I

This method for reducing or preventing human error-related anesthetic mishaps calls for appropriate actions in the following five areas [11]:

- **Organization**. Two actions called for in this area are to (1) improve communication between all staff members and (2) revise work policies to reduce haste or stress.
- **Education**. Two actions called for in this area are to (1) review all reported incidents on a regular basis and (2) upgrade all operating room staff knowledge through courses and seminars.
- **Equipment**. Three actions called for in this area are to (1) adopt appropriate measures to check equipment prior to its use, (2) provide

adequate number and types of monitors, and (3) replace faulty or inappropriate pieces.

- **Supervision.** Two actions called for in this area are to (1) ensure complete preoperative assessment of patients and (2) ensure available supervision and additional help if required.
- **Protocols.** One action called for in this area is to formulate protocols for repetitive tasks, patient monitoring, and patient transport.

8.2.1.2 Method II

This method for reducing or preventing human error-related anesthetic mishaps is composed of six steps, as shown in Figure 8.1 [12]. Step 1 is concerned with determining the existing state of affairs by reviewing items such as hazard alerts, medical defense reports, incident monitoring studies, mortality committee reports, observation studies, morbidity committee reports, and simulation studies.

Step 2 is concerned with collating information and collecting subsets for studies and classifications. This calls for establishing a project team that has adequate access to the appropriate information. Step 3 is concerned with the grouping of problems, errors, and contributory factors under appropriate classifications.

Step 4 is concerned with developing appropriate strategies with respect to detecting, preventing, avoiding, and minimizing the potential consequences. Step 5 is concerned with the implementation of strategies at appropriate places. Effective communication plays an important role here. Finally, Step 6 is concerned with assessing the effectiveness of strategies in action and enforcing appropriate corrective measures (if necessary).

8.3 Human Error in Emergency Medicine

Just like any other area of the health care system, emergency departments are subject to human error. There are about 100 million emergency department patient visits in the United States each year [13]. Thus, even a very small percentage of human error occurrence in this area can translate into a considerable number of associated adverse events. Some of the facts and figures that directly or indirectly support the occurrence of human errors in emergency departments are as follows:

- Over 90% of the adverse events in emergency departments are preventable [1,3].

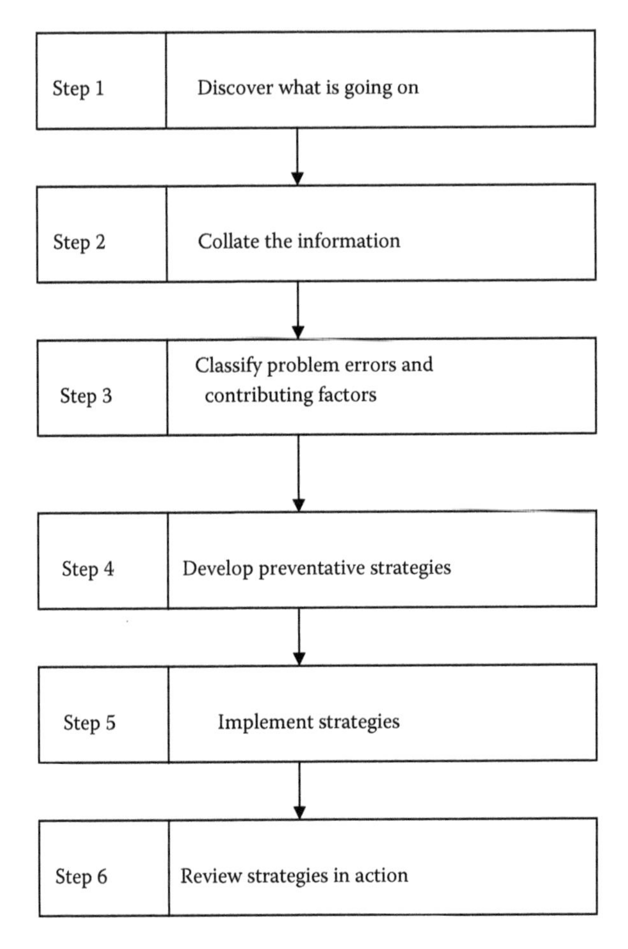

FIGURE 8.1
Method II steps.

- A study of missed diagnoses of acute cardiac ischemia in an emergency department reported that 4.3% of 1817 patients with acute cardiac ischemia were mistakenly discharged [13].
- A study of error rate in emergency physician interpretation of the cause in electrocardiographic (ECG) ST-Segment evaluation (STE) of 202 adult chest pain patients who had STEs reported that the rate of ECG STE misinterpretation was about 6% [14].
- A study of the interpretation of radiographs reported that the rates of disagreement between radiologists and emergency physicians vary from 8% to 11% [15].

To prevent or minimize human error occurrences in emergency medicine, it is essential to accurately assess all possible risks or predictor factors.

Therefore, from the public health perspective, appropriate epidemiological approaches should be used to identify the causes of emergency medicine-related errors and their resulting adverse events or consequences [13,16,17]. Also, questions such as those presented below can be useful in reducing the occurrence of human errors in the area of emergency medicine [13]:

- Are there any ideal length of shifts and change-of-shift methods to reduce human errors in emergency medicine?
- What are the possible effects of reducing the distractions (e.g., telephone calls and paging interruptions) of health care providers on the occurrence of human errors in emergency medicine?
- Is it possible to make the occurrences of human errors more visible?
- Is the presence of a pharmacologist helpful in reducing the occurrence of adverse events and human errors in the emergency department setting?
- Are the computerized clinical information systems useful to reduce the occurrence of adverse events and human errors in the emergency department setting?

8.4 Human Error in Operating Rooms

Professionals such as surgeons, anesthetists, surgical nurses, anesthetic nurses, and individuals from the support services form the operating room teams [18]. These teams usually work under time pressure, and the effectiveness of team performance in the operating room environment is very important, as surgical tasks require the coordinated efforts of all involved individuals. The occurrence of a human error in such an environment can result in patient death or permanent damage to a patient.

Some examples of the human errors that can occur in operating rooms are administering an inappropriate drug, severing an artery, and failing to note falling blood pressure. Human errors observed in operating rooms from the behavioral aspect can be grouped under three classifications as shown in Figure 8.2 [18]. These classifications are communications/decisions, preparation/planning/vigilance, and workload distribution/distraction avoidance.

Some examples of the human errors that belong to the communication/decisions classification are the failure of the surgeon to inform the anesthetist of drugs having effects on blood pressure, the consultant drawing up patient schedules without properly informing the resident or the nursing staff, and the consultant leaving work-overloaded resident staff in the operating room. Three examples of the human errors that belong to the

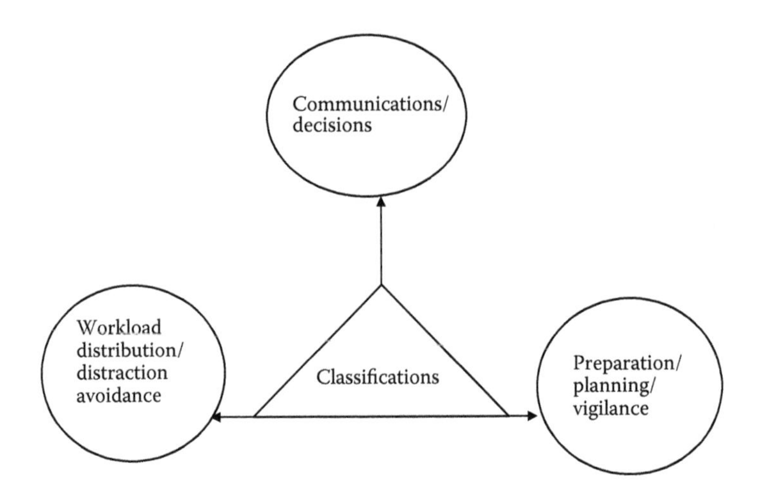

FIGURE 8.2
Classifications of human errors in operating rooms from the behavioral aspect.

preparation/planning/vigilance classification are failure to react to blood pressure and blood oxygen alarms, failure to monitor patient status during operation, and failure to complete checklist (e.g., anesthesia machine set incorrectly).

Finally, two examples of the human errors that belong to the workload distribution/distraction avoidance classification are the resident staff reading the technical manual, resulting in the patient not being adequately relaxed, and the consultant being distracted from making a decision to place a pulmonary artery catheter by problems identified by another operating facility.

To study the occurrence of human errors in operating room environments effectively, it is important to understand the model representing operating room performance. The model is composed of the following four main elements [18]:

- **Team input factors.** These factors include items such as team composition, time pressure, patient condition, individual aptitudes, organizational climate/norms, and personality/motivation.
- **Team performance functions.** Some of the items associated with team performance functions are team management, communication skills, team formation, situational awareness, decision processes, conflict resolution, and technical procedures.
- **Individual and organizational outcomes.** Three items associated with individual and organizational outcomes are attitudes, professional development, and morale.
- **Team outcomes.** Two items associated with team outcomes are patient safety and team efficiency.

Studies of this model indicate the fact that there are numerous instances that can lead to the occurrence of human error in the operating room environment. Some of the observed problems associated with the operating team coordination are listed below [2].

- Unresolved differences between anesthetists and surgical team members.
- Failure to establish effective leadership for the operating team.
- Failure to effectively brief one's own team and the other teams during the operation planning process.
- Consultants' failure to provide satisfactory training to all involved residents.
- Failure to discuss alternative procedures and advocate own position clearly as well as to inform all team members about patient or workload-related problems.
- Failure to debrief operation effectively to learn from past experiences for future reference.
- Frustration because of unsatisfactory team coordination.

8.5 Human Error in Intensive Care Units

Each day thousands of patients throughout the United States are warded into intensive care units where they receive various types of medical services. Just as in the performance of any other services in the health care system, the performance of services in intensive care units is subjected to human error. Some of the facts and figures concerned with the occurrence of human error in intensive care units are presented below:

- A study of critical incidents that occurred over a 10-year period (1989–1999) in an intensive care unit reported that most of these incidents were the result of staff errors [4].
- A study of seven intensive care units performed over a 1-year period reported that there were 610 incidents in these units. A detailed analysis of these incidents revealed that 66% of them were due to human factor-related problems and the remaining 34% were due to system-related problems [19].
- A study of an 11-bed multidisciplinary intensive care unit performed over 1 year reported that 241 errors occurred in the unit during the specified period. A detailed analysis of these human errors revealed

that one human error was lethal, two resulted in sequelae, 26% led to prolonged stay in the intensive care unit, 57% were minor, and 16% were free of any consequence [20].

- A study of 145 adverse events that occurred during the period from 1974 to 1978 involving patients in an intensive care unit reported that 92 of these adverse events were due to human error [21].
- A study of a 6-bed intensive care unit performed over a period of 6 months reported that there were 554 human errors [22]. A detailed analysis of these human errors revealed that about 29% of them were considered severe or potentially detrimental to the patients. Furthermore, about 55% of the errors were committed by nurses and the remaining 45% by physicians [22].

Various studies concerned with the analysis of human errors in intensive care units have identified the following four basic factors for the occurrence of human error in intensive care units [4]:

- Poor communication
- Inadequate experience
- Staff shortage
- Night time

In regard to the detection of incidents in intensive care units, some of the important contributing factors identified by these studies are the presence of properly experienced staff members, regular checking, and the presence of working alarms on the equipment used.

8.6 Human Error in Medical Technology Use, Laboratory Testing, Radiotherapy, and Image Interpretation

With the increase in complexity and sophistication of medical technology, the occurrence of human errors in medical technology use has increased quite significantly [23,24]. One study reported that over 50% of medical device failures were due to operators, actions taken by involved patients, maintenance, and service [25]. Some important factors for the occurrence of human errors in medical technology are as follows [2]:

- Inadequate or no consideration to human factors during device design
- Poorly written device operating and maintenance procedures

- Inadequate or no training at all in device use
- The poor standardization of devices

Various studies have reported the occurrence of human errors in the area of laboratory testing [2,23]. For example, as per New York State laboratory regulators there were around 66% testing errors in the laboratories offering drug-screening services [23,26]. This shows that there is a definite need to improve the laboratory testing system to reduce the occurrence of human errors.

Radiotherapy is also prone to the occurrence of human errors. More specifically, in the use of nuclear materials for patient therapy there are numerous opportunities for the occurrence of human errors. Two examples of human errors in this area are as follows [2,23]:

- A patient was accidentally administered 100 rad of radioactive materials to the brain.
- A patient was administered 50 millicuries of radioactive materials instead of 3 millicuries as recommended by the physician.

Many studies have reported the occurrence of human errors in the interpretation of medical results from x-rays, electrocardiograms, CAT scans, and sonograms [23,27]. The image misinterpretation problem is often associated with junior and inexperienced doctors [23,27,28]. Additional information on this topic is available in Refs. [23,29].

8.7 Factors Contributing to Human Error in Surgical Pathology and Causes of Wrong-Site Surgeries

There are many factors that contribute to human error in surgical pathology. Some of these factors are shown in Figure 8.3 [30] and include complexity, time constraints, inconsistency, variable input, inflexible hierarchical culture, human intervention, and hands-off. Additional information on each of these factors is available in Ref. [30].

Wrong-site surgery (WSS) has become an important issue, with an increasing number of cases reported every year [31]. WSS encompasses surgery performed on the wrong site or side of the patient's body, surgery performed on the wrong patient, and incorrect surgical procedure performed [32]. Cases of WSS frequently occur in general surgery, urological and neurosurgical procedures, and orthopedic or podiatric procedures [31].

The causes of WSSs may be divided under the following two broad categories [31,33]:

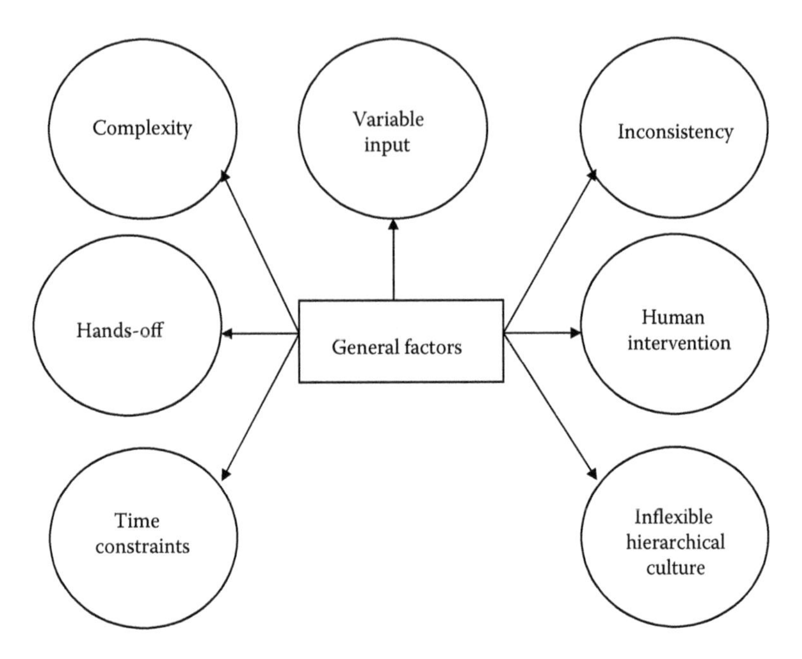

FIGURE 8.3
General factors contributing to human error in surgical pathology.

- System factors
- Process factors

The *system factors* category includes causes such as lack of institutional controls/formal system to verify the correct site surgery, pressures to reduce preoperative preparation time, lack of a checklist to make sure every check was performed, work environment, unusual time pressures (e.g., unplanned emergencies or large volume of procedures), reliance solely on the surgeon for determining the correct surgical site, team competency and credentialing, exclusion of certain surgical team members, availability of information, training and orientation, staffing, continuum of care, patient characteristics (e.g., obesity or unusual anatomy) that require alterations in the usual positioning of the patient, and procedures requiring unusual equipment or patient positioning [33].

The *process factors* category includes causes such as poor care planning, more than one surgeon involved in the procedure, inadequate patient assessment, noncompliance with procedures, failure to recheck patient information prior to starting the operation, multiple procedures on multiple parts of a patient performed during a single operation, inadequate medical record review, failure to mark or clearly mark the correct operation site, miscommunication among members of the surgical team and the patient, and inaccurate or incomplete communication among members of the surgical team [33].

8.8 Reasons for Clinicians Not Reporting and Disclosing Errors and Near Misses

Various studies have been performed to determine the reasons for clinicians not reporting and disclosing errors and near misses [34]. The reasons found may be divided under four main categories, as shown in Figure 8.4 [34]. These categories are fear, burden of effort, understanding, and administrative/management/organizational.

The *fear* category includes reasons such as fear of being blamed for negative patient outcome, fear of reprimand from physician(s), fear of "telling" on someone else, fear that patients will develop negative attitudes, fear of adverse consequences from reporting, fear of legal liability, fear of reporting that is not anonymous, and fear that other providers will consider the provider who made the error incompetent [34–43].

The *burden of effort* category includes reasons such as incident reports taking too long to complete, extra work involved in reporting, and verbal reports to physicians taking too long or contacting the doctor takes too much time [38–40,43–45].

Some of the main reasons belonging to the *understanding* category are confusion over definitions of near misses and errors, no perceived benefits, disagreement with the organization's definition of error, providers' bias about which incidents should be reported, providers unaware that errors occurred, and some incidents (i.e., near misses) thought too trivial to report [38–40,42,43,47].

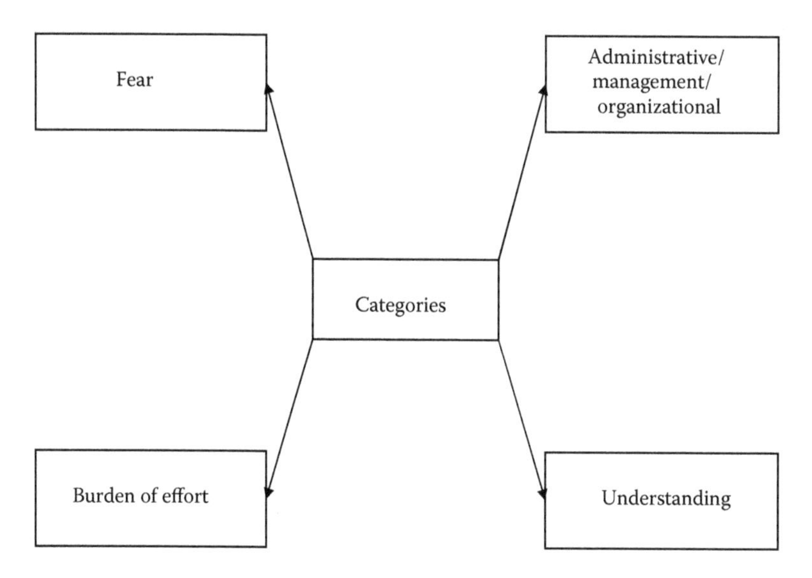

FIGURE 8.4
Main categories of reasons for clinicians not to report and disclose errors and near misses.

Finally, the *administrative/management/organizational* category includes reasons such as lack of feedback on reported errors, administrative response, poor match of administrative response to errors with severity of errors, and persistence of the culture of blame/shame, blaming the individual [37–39,42,46].

8.9 Guidelines for Preventing the Occurrence of Medical Errors

Various guidelines have been developed to prevent the occurrence of medical errors. Ten of these guidelines are as follows [48]:

- **Simplify.** This guideline calls for reducing the number of steps involved in a work process, the number of times an instruction is given, the use of nonessential equipment, software, and procedures, and so on.
- **Standardize.** This guideline calls for limiting the unnecessary variety in equipment, rules, drugs, and so on.
- **Make improvements in communication patterns.** This guideline calls for team members in areas such as intensive care units, emergency departments, and operating rooms to repeat all orders to make sure that they have understood them clearly and correctly.
- **Use sensible checklists.** This guideline calls for developing and using checklists sensibly and effectively. Although checklists and procedures help to minimize variables as well as provide a greater possibility of having consistent results, when they become something other than tools, they can take away human judgment.
- **Stratify.** This guideline is concerned with over-standardization. It warns against over-standardization (i.e., one size fits all) because it can cause various types of errors.
- **Make use of affordances.** This guideline calls for designing features in items that ensure correct use by providing appropriate clues to proper operation.
- **Redesign the patient record for effectiveness.** This guideline calls for examining the effectiveness of the existing form of records keeping and then taking necessary actions.
- **Automate cautiously.** This guideline warns against over-automation, which can prevent operators or others from judging the true state of the system under consideration.

- **Use defaults effectively.** This guideline calls for making the correct action the easiest one. (A default is a standard rule/order that works if nothing else intervenes.)
- **Respect human shortcomings.** This guideline calls for giving proper consideration to human-related factors such as stress, memory limitations, time pressure, workload, and circadian rhythm in designing work systems and tasks.

Additional information on the above guidelines is available in Refs. [2,48].

8.10 Health Care Human Error Reporting Systems

Human error data play an important role in making various types of decisions in the health care sector. The effectiveness of such decisions depends on the error data quality (i.e., poor-quality data will result in ineffective decisions). There are many human-error-related systems currently in use in the area of health care [2,5], some of which are described below [2,5]:

- **Joint Commission on Accreditation of Healthcare Organizations (JCAHO) Event Reporting System.** This system was introduced by the JCAHO in 1996 and is a sentinel event reporting system for hospitals. A sentinel event is defined as an unexpected variation or occurrence that involves serious physical or psychological injury/death/the risk thereof [5]. An organization experiencing a sentinel event is required to conduct root cause analysis by JCAHO to determine event causal factors. In regard to the reporting of the sentinel event to JCAHO, a hospital may voluntarily report an incident and then submit its associated root cause analysis accompanying the proposed measures for improvement.
- **Food and Drug Administration (FDA) Surveillance System.** This was developed and is managed by the FDA. As part of this system, all adverse events reports concerning medical products after their formal approval are submitted to the FDA. In the case of medical devices, the device manufacturers report information on items such as deaths, serious injuries, and malfunctions. Furthermore, the device user facilities such as hospitals and nursing homes are required to report deaths to both the FDA and manufacturers and also serious injuries to device manufacturers. In regard to drug-related adverse events, reporting is mandatory for all manufacturers, but it is voluntary for physicians, consumers, and so on [2].

- **Medication Errors Reporting System.** This system was developed by the Institute for Safe Medication Practice (ISMP) in 1975 and is a voluntary medication error reporting system. The system receives reports from frontline practitioners and shares its information with the pharmaceutical companies and the FDA, and is managed by U.S. Pharmacopoeia (USP).
- **Med Marx System.** This Internet-based system is established for hospitals to report medication errors anonymously on a voluntary basis. The system was developed by the USP in 1998 and the information contained in the system is not shared with the FDA.
- **State Adverse Event Tracking Systems.** These systems are used by various U.S. state governments to monitor adverse events occurring in health care organizations. Two main impediments associated with these systems are the limitations in data and the lack of required resources. Additional information on these systems is available in Refs. [2,5].

8.11 Problems

1. List at least 12 common anesthesia errors.
2. Describe two methods for reducing or preventing human error-related anesthetic mishaps.
3. Write an essay on human error in emergency medicine.
4. What are the classifications of human errors in operating rooms from the behavioral aspect?
5. What are the observed problems associated with the operating team coordination?
6. Discuss human error in intensive care units.
7. What are the general factors that contribute to the occurrence of human error in surgical pathology?
8. Discuss human error in medical technology use and laboratory testing.
9. What are the main categories of reasons for clinicians not to report and disclose errors and near misses?
10. List and discuss at least seven useful guidelines for preventing the occurrence of medical errors.

8.12 References

1. Wears, R. L., Leape, L. L., Human Error in Emergency Medicine, *Annals of Emergency Medicine*, Vol. 34, No. 3, 1999, pp. 370–372.
2. Dhillon, B. S., *Human Reliability and Error in Medical System*, World Scientific, River Edge, New Jersey, 2003.
3. Bogner, M. S., Editor, *Human Error in Medicine*, Lawrence Erlbaum Associates, Hillsdale, New Jersey, 1994.
4. Wright, D., *Critical Incident Reporting in an Intensive Care Unit*, Report, Western General Hospital, Edinburgh, Scotland, UK, 1999.
5. Kohn, L. T., Corrigan, J. M., Donaldson, M. S., Editors, *To Err Is Human: Building a Safer Health System*, Institute of Medicine, National Academy Press, Washington, D.C., 1999.
6. Bates, D. W., et al., The Costs of Adverse Drug Events in Hospitalized Patients, *JAMA*, Vol. 277, No. 4, 1997, pp. 307–311.
7. Cooper, J. B., et al., Preventable Anesthesia Mishaps, *Anesthesiology*, Vol. 49, 1978, pp. 399–406.
8. Edwards, G., Morton, H. J. V., Pask, E. A., Deaths Associated with Anesthesia: Report on 1000 Cases, *Anesthesia*, Vol. 11, 1956, pp. 194–220.
9. Cooper, J. B., Toward Prevention Anaesthetic Mishaps, *International Anesthesiology Clinics*, Vol. 22, 1984, pp. 167–183.
10. Craig, J., Wilson, M. E., A Survey of Anaesthetic Misadventures, *Anesthesia*, Vol. 36, 1981, pp. 933–938.
11. Short, T. G., et al., Critical Incident Reporting in an Anesthetic Department Quality Assurance Program, *Anesthesia*, Vol. 47, 1992, pp. 3–7.
12. Runciman, W. B., et al., Errors, Incidents, and Accidents in Anesthetic Practice, *Anesthesia and Intensive Care*, Vol. 21, No. 5, 1993, pp. 506–518.
13. Pope, J. H., et al., Missed Diagnoses of Acute Cardiac Ischemia in the Emergency Department, *New England Journal of Medicine*, Vol. 342, No. 16, 2000, pp. 1163–1170.
14. Brady, W. J., Perron, A., Ullman, E., Errors in Emergency Physician Interpretation of ST-Segment Elevation in Emergency Department Chest Pain Patients, *Academic Emergency Medicine*, Vol. 7, No. 11, 2000, pp. 1256–1260.
15. Espinosa, J. A., Nolan, T. W., Reducing Errors Made by Emergency Physicians in Interpreting Radiographs: Longitudinal Study, *British Medical Journal*, Vol. 320, 2000, pp. 737–740.
16. Rothman, K. J., Lanes, S., Robins, J., Casual Inference, *Epidemiology*, Vol. 4, 1993, pp. 555–556.
17. Robertson, L., *Injury Epidemiology: Research and Control Strategies*, Oxford University Press, New York, 1998.
18. Helmreich, R. L., Schaefer, H. G., Team Performance in the Operating Room, in *Human Error in Medicine*, edited by M. S. Bogner, Lawrence Erlbaum Associates, Hillsdale, New Jersey, 1994, pp. 225–253.
19. Beckmann, V., et al., The Australian Incident Monitoring Study in Intensive Care (AIMS-ICU): An Analysis of the First Year of Reporting, *Anesthesia and Intensive Care*, Vol. 24, No. 3, 1996, pp. 320–329.
20. Bracco, D., et al., Human in a Multidisciplinary Intensive Care Unit: A 1-Year Prospective Study, *Intensive Care Medicine*, Vol. 27, 2001, pp. 137–145.

21. Abramson, N. S., et al., Adverse Occurrences in Intensive Care Units, *JAMA*, Vol. 244, No. 14, 1980, pp. 1582–1584.
22. Donchin, Y., et al., A Look into the Nature and Causes of Human Errors in the Intensive Care Unit, *Critical Care Medicine*, Vol. 23, No. 2, 1995, pp. 294–300.
23. VanCott, H., Human Errors: Their Causes and Reduction, in *Human Error in Medicine*, edited by M. S. Bogner, Lawrence Erlbaum Associates, Hillsdale, New Jersey, 1994, pp. 53–65.
24. Cooper, J. B., et al., An Analysis of Major Errors and Equipment Failures in Anesthesia Management: Considerations for Prevention and Detection, *Anesthesiology*, Vol. 60, 1984, pp. 34–41.
25. Nobel, J. L., Medical Device Failures and Adverse Effects, *Pediatric Emergency Care*, Vol. 7, 1991, pp. 120–123.
26. Squires, S., Cholesterol Guessing Games, *Washington Post*, Washington, D.C., March 6, 1990.
27. Morrison, W. G., Swann, I. J., Electrocardiograph Interpretation by Junior Doctors, *Archives of Emergency Medicine*, Vol. 7, 1990, pp. 108–110.
28. Vincent, C. A., et al., Accuracy of Detection of Radiographic Abnormalities by Junior Doctors, *Archives of Emergency Medicine*, Vol. 5, 1988, pp. 101–109.
29. Rolland, J. P., Barrett, H. H., Seeley, G. W., Ideal versus Human Observer for Long Tailed Point Spread Functions: Does De-convolution Help?, *Physiology, Medicine, Biology*, Vol. 36, No. 8, 1991, pp. 1091–1109.
30. Nakhleh, R. E., Patient Safety and Error Reduction in Surgical Pathology, *Archives of Pathology & Laboratory Medicine*, Vol. 132, February 2008, pp. 181–185.
31. Mulloy, D. F., Hughes, R. G., Wrong-Site Surgery: A Preventable Medical Error, in *Patient Safety and Quality: An Evidence-Based Handbook for Nurses*, edited by R. G. Hughes, Agency for Healthcare Research and Quality, Rockville, Maryland, 2008, Chapter 36, pp. 1–11.
32. Carayon, P., Schultz, K., Hundt, A. S., Righting Wrong Site Surgery, *Jt. Comm. J. Qual. Saf.*, Vol. 30, 2004, pp. 405–410.
33. Saufl, N. M., Universal Protocol for Preventing Wrong Site, Wrong Procedure, Wrong Person Surgery, *Journal of PeriAnesthesia Nursing*, Vol. 19, 2004, pp. 348–351.
34. Wolf, Z. R., Hughes, R. G., Error Reporting and Disclosure, in *Patient Safety and Quality: An Evidence-Based Handbook for Nurses*, edited by R. G. Hughes, Agency for Healthcare Research and Quality, Rockville, Maryland, 2008, Chapter 35, pp. 1–47.
35. Osborne, J., Blais, K., Hayes, J. S., Nurses' Perceptions: When Is It a Medication Error?, *Journal of Nursing Administration*, Vol. 29, No. 4, 1999, pp. 33–38.
36. Blegen, M. A., Vaughn, T., Pepper, G. et al., Patient and Staff Safety: Voluntary Reporting, *American Journal of Medical Quality*, Vol. 19, No. 2, 2004, pp. 67–74.
37. Wakefield, D. S., et al., Understanding and Comparing Differences in Reported Medication Administration Error Rates, *American Journal of Medical Quality*, Vol. 14, No. 2, 1999, pp. 73–80.
38. Wakefield, B. J., et al., Organizational Culture, Continuous Quality Improvement, and Medication Error Reporting, *American Journal of Medical Quality*, Vol. 16, No. 4, 2001, pp. 128–134.
39. Chiang, H., Pepper, G. A., Barriers to Nurses' Reporting of Medication Administration Errors in Taiwan, *Journal of Nursing Scholarship*, Vol. 38, No. 4, 2006, pp. 392–399.

40. Wakefield, D. S., et al., Perceived Barriers in Reporting Medications Administration Errors, *Best Practices & Benchmarking in Healthcare*, Vol. 1, No. 4, 1996, pp. 191–197.
41. Stratton, K. M., et al., Reporting of Medication Errors by Pediatric Nurses, *Journal of Pediatric Nursing*, Vol. 19, No. 6, 2004, pp. 385–392.
42. Mayo, A. M., Duncan, D., Nurse Perceptions of Medication Errors: What We Need to Know for Patient Safety, *Journal of Nursing Care Quality*, Vol. 19, No. 3, 2004, pp. 209–217.
43. Uribe, C. L., et al., Perceived Barriers to Medical-Error Reporting: An Exploratory Investigation, *Journal of Health Care Management*, Vol. 47, No. 4, 2002, pp. 263–280.
44. Elder, N. C., et al., Barriers and Motivators for Making Error Reports from Family Medicine Offices: A Report from the American Academy of Family Physicians National Network (AAFP NRN), *Journal of the American Board of Family Medicine*, Vol. 20, 2007, pp. 115–123.
45. Evans, S. M., et al., Attitudes and Barriers to Incident Reporting: A Collaborative Hospital Study, *Quality & Safety in Health Care*, Vol. 15, 2006, pp. 39–43.
46. Evans, S. M., et al., Evaluation of an Intervention Aimed at Improving Voluntary Incident Reporting in Hospitals, *Quality & Safety in Health Care*, Vol. 16, 2007, pp. 169–175.
47. King, G., Perceptions of Intentional Wrong-Doing and Peer Reporting Behavior among Registered Nurses, *Journal of Business Ethics*, Vol. 34, 2001, pp. 1–13.
48. Crane, M., How Good Doctors Can Avoid Bad Errors, *Medical Economics*, April 1997, pp. 36–43.

9

Medical Device Safety and Errors

9.1 Introduction

A medical device is expected to be not only reliable but also safe. This means that a given medical device must never function or malfunction in a manner that can be harmful to the user or the patient. The problem of safety in regard to humans is not new; its history can be traced back to the ancient Babylonian ruler Hammurabi, who developed a code known as the "Code of Hammurabi" in 2000 B.C. [1–3]. The code contained clauses concerning injury and financial damages against those individuals who cause injury to other people.

In modern times, in 1970 the U.S. Congress passed the Occupational Safety and Health Act (OSHA), which is considered an important milestone in regard to health and safety in the United States. Two subsequent milestones specifically concerned with medical devices in the United States are the Safe Medical Device Act of 1990 and the Medical Device Amendments in 1976.

The occurrence of human errors in the area of medical devices has become an important issue, as operator errors account for over 50% of all technical medical equipment problems [4]. Furthermore, as per Refs. [5–7], human error is considered to contribute up to 90% of accidents both generally and in the area of medical devices.

This chapter presents various important aspects of medical device safety and errors.

9.2 Types of Medical Device Safety and Medical Device Hardware and Software Safety

Medical device safety may be classified under the following three categories [8]:

- **Unconditional safety.** This means safety without any condition whatsoever. This type of safety is preferred over all other types because it is the most effective. However, it demands eradication of all types of risks associated with medical devices by design.

- **Conditional safety.** This type of safety is used when it is not possible to realize unconditional safety. For example, when x-rays or laser surgical devices are used, it is impossible to avoid dangerous radiation emissions. However, it is possible to minimize risk through measures such as limiting access to therapy rooms or incorporating a locking switch that allows device activation by authorized personnel only. Other, indirect safety means are protective laser glasses, x-ray folding screens, and so on.
- **Descriptive safety.** This type of safety is used in situations when it is not possible to provide safety by unconditional or conditional means (i.e., the above two means). Descriptive safety in regard to operation, mounting, replacement, transport, connection, and maintenance may simply be statements such as "Handle with care," "This side up," and "Not for explosive zones."

The hardware safety of medical devices is important because many of their parts, such as electronic parts, are vulnerable to factors such as electrical interferences and environmental stresses. This calls for analysis of each part of a medical device with respect to safety and potential failures. There are several methods that can be used to perform such analysis, two of which are failure modes and effect analysis (FMEA) and fault tree analysis (FTA).

Subsequent to parts analysis, to reduce the potential for the failure of parts/components identified as critical, approaches such as safety margin, load protection, and component derating can be used.

Safety of software is equal in importance to that of the hardware elements, as consequences of a software failure in medical devices can be quite serious. For example, a program that is out of control because of a software error can drive a radiation therapy machine gantry into a patient [9]. A Food and Drug Administration (FDA) study conducted over six years (1983–1989) reported that there were 116 problems in software quality that resulted in the recall of medical devices in the United States [3]. Most of the methods and techniques that can be used to improve software safety in medical devices are available in Refs. [10–12].

9.3 Essential Safety Requirements for Medical Devices, Legal Aspects of Medical Device Safety, and Medical Device Electrical Safety Standards

There are various requirements placed by the government and other agencies on medical devices with respect to safety. They may be grouped under the following three areas [8]:

- **Safe design.** The requirements that belong to this area include protection against radiation hazards, protection against electrical shock, mechanical hazard prevention, care for hygienic factors, excessive heating prevention, care for environmental conditions, and proper material choice with respect to chemical, mechanical, and biological factors. Although all these requirements are considered self-explanatory, the requirements care for environmental conditions and mechanical hazard prevention are described further. The care for the environmental conditions requirement includes factors such as temperature, humidity, and electromagnetic interactions. The mechanical hazard prevention requirement includes factors such as safe distances, breaking strength, and device stability.
- **Safe function.** The requirements that belong to this area include accuracy of measurements, warning for or prevention of dangerous outputs, and reliability.
- **Sufficient information.** The requirements that belong to this area include instructions for use, effective labeling, and accompanying documentation.

The legal aspects of medical device safety have become an important issue. Three commonly used theories to make manufacturers liable for injury caused by their product are shown in Figure 9.1 [3,13]. These are breach of warranty, negligence, and strict liability. Breach of warranty may be alleged under three scenarios: breach of an expressed warranty, breach of the implied warranty of merchantability, and breach of the implied warranty of suitability for a specific case. For example, if a medical device causes an injury to a person because of its failure to operate as warranted, the device manufacturer faces liability under the expressed warranty scenario.

In the case of negligence, if a medical device manufacturer fails to exercise reasonable care or fails to meet a reasonable standard of care during

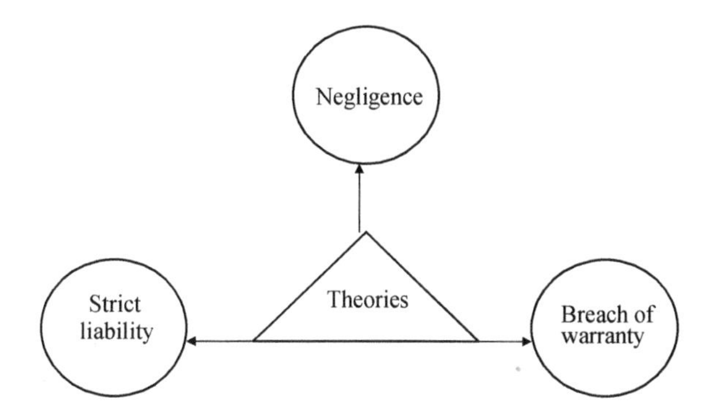

FIGURE 9.1
Theories to make manufacturers liable for injury caused by their products.

manufacture, handling, or distribution of the device, then it could be liable for any damages caused by the device. Finally, in the case of strict liability, the basis for its imposition is that the manufacturer of the device is in the best position to reduce all types of associated risks.

Product safety certification agencies use various types of safety standards to evaluate engineering products. These standards are documents that clearly define the products' minimum construction and performance requirements with respect to safety. All types of electrically operated medical equipment/devices used in the United States are evaluated according to standard UL 2601-1 (IEC 60601-1) [14–17]. The UL 2601-1 safety standard contains the full text of IEC 60601-1 in addition to the U.S. deviations [14]. The U.S. deviations contain items such as the following [14]:

- Component requirements
- Enclosure flame ratings
- Production line testing
- Lower leakage current limits

The main objective of using an internationally coordinated safety standard such as IEC 60601-1/UL2601-1 is that a medical device could be designed and evaluated for compliance with a single standard and thus can be sold in many different countries without any problem. Currently, the countries that use the IEC 60601-1 standard include Canada, European Union countries, Japan, Australia, and South Korea.

Additional information on medical device electrical safety standards is available in Refs. [14–17].

9.4 Software-Related Issues in the Safety of Cardiac Rhythm Management Products

There are various software-related issues in the safety of cardiac rhythm-management products (e.g., pacemakers) that must be considered with care when analyzing their software safety. These issues may be divided under the following three areas [18]:

- Marketing issues
- Technical issues
- Management issues

The marketing issues are made up of four main components: market requirements, regulatory requirements, legal requirements, and product requirements. In the case of market requirements, the sheer size of the cardiac rhythm management products/systems market is the sole important factor in regard to their safety. For example, in the United States about half a million people each year experience a sudden cardiac death episode and over 30,000 individuals receive a defibrillator implant [19].

The regulatory requirements are important because regulatory agencies such as the Food and Drug Administration (FDA) require systematic and rigorous software development processes concerning cardiac rhythm management systems/products, including safety analysis. The legal requirements are another important component of the marketing issues because of the life-or-death nature of cardiac rhythm management systems/products. The regulations concerning these systems/products can lead to highly sensitive legal requirements involving patients and their families, regulatory agencies, manufacturers, and so on.

Finally, in the case of product requirements, a typical cardiac rhythm management device is composed of items such as advanced software, mechanical subsystems, and electrical subsystems, which must function effectively for the device's overall success. In particular, as the modern cardiac rhythm management devices' software subsystem is generally made of around half a million lines of code, its reliability, safety, and efficiency for controlling both internal and external operations are very important.

Technical issues are an important factor during the software development process because software complexity and maintenance-related concerns determine the analysis technique and the incorporation process to be employed.

Finally, management issues are concerned with making changes to the ongoing software development process for incorporating explicit safety analysis, thus requiring very convincing justifications and clear vision. More specifically, the management personnel have to be convinced that additional tasks concerning safety will help the established goals in the long run despite the perceived extra short-term efforts.

Additional information on management and technical issues is available in Refs. [3,18].

9.5 Classifications of Medical Device Accidents and Medical Device Accident Occurrence Probability Estimation

There are various types of medical device accidents. They may be classified under the following seven categories [20]:

- Random component failure
- Operator or patient error

- Faulty calibration, preventive maintenance, or repair
- Abnormal or idiosyncratic patient response
- Design deficiency
- Manufacturing defect
- Sabotage or malicious intent

The probability of an accident occurrence from the operation of a medical device can be estimated by using the following equation [21]:

$$P_{mda} = A + B + C \tag{9.1}$$

$$A = \sum P_{ih}(1 + P_{ac} + P_{ae}) + P_{rh}(1 + P_{ac} + P_{ae})(1 - P_{ce}) \tag{9.2}$$

$$B = \sum P_{fn}(1 + P_{fm} + P_{ef}) + P_{ft}(1 + P_{fm} + P_{ef})(1 - P_{ce}) \tag{9.3}$$

$$C = \sum (P_{ai} + P_{ei})(1 - P_{ce}) \tag{9.4}$$

where

P_{mda} = the probability of an accident occurrence from the operation of a medical device.

P_{ac} = the probability of the device having an adverse characteristic that can result in human error/failure.

P_{rh} = the occurrence probability of a repairable human error that can cause or allow an accident. Some examples of such human error are an incorrect action, an unsatisfactory response, and a wrong decision.

P_{ih} = the occurrence probability of an irreparable human error that could cause or allow an accident.

P_{ce} = the probability of correct action taken as per requirement (e.g., correct corrective measure, satisfactory response, and correct decision).

P_{ae} = the probability of the device encountering an extraordinary and adverse environment that can result in human error.

P_{ef} = the probability of the device encountering an extraordinary and adverse environment that can result in device failure.

P_{fm} = the probability of the device having an adverse characteristic that can result in material failure.

P_{ft} = the occurrence probability of those failures that will result in accidents unless possible corrective actions are taken in a timely manner.

P_{fn} = the occurrence probability of those failures that will result in accidents to which no corrective action is possible.
P_{ei} = the probability of the device encountering an extraordinary and adverse environment that could result in injury or damage without the occurrence of error or failure.
P_{ai} = the probability of the device having an adverse characteristic that can result in loss, damage, or injury without the occurrence of material failure or error.

9.6 Medical Devices with a High Incidence of Human Error

The occurrence of human errors in using medical devices is a pressing problem. For example, human errors in using medical devices cause, on average, at least three serious injuries or deaths per day [22]. Therefore, over the years various types of studies have been conducted to identify those medical devices that have a high incidence of human error. Such studies revealed data on the following devices, listed in the order of most error-prone to least error-prone [22]:

- Glucose meter
- Balloon catheter
- Orthodontic bracket aligner
- Administration kit for peritoneal dialysis
- Permanent pacemaker electrode
- Implantable spinal cord simulator
- Intravascular catheter
- Infusion pump
- Urological catheter
- Electrosurgical cutting and coagulation device
- Nonpowered suction apparatus
- Hydraulic/mechanical impotence device
- Implantable pacemaker
- Peritoneal dialysate delivery system
- Catheter introducer
- Catheter guide wire
- Transluminal coronary angioplasty catheter
- External low-energy defibrillator

- Continuous ventilator (respirator)
- Contact lens cleaning and disinfecting solutions

9.7 Medical Device Operator Errors

Many operator-related errors occur during the operation or maintenance of medical devices/equipment. The chief ones are as follows [23]:

- Incorrect decision making and measures to critical situations
- Misinterpretation of a failure to recognize important device outputs
- Errors in setting device parameters
- Inadvertent/untimely/inappropriate activation of device controls
- Failure to follow specified instructions and procedures
- Overreliance on a medical device's automatic features, alarms, or capabilities
- Wrong selection of devices in regard to the clinical requirements and objectives
- Misassembly
- Inappropriate improvisation

9.8 General Approach to Human Factors during the Medical Device Development Process for Reducing Human Errors

Human errors can be reduced substantially by making human factors an integral part of the medical device/equipment development process (i.e., right from the concept phase to the production phase) as shown in Figure 9.2 [24].

During the concept phase, the human factors specialist performs tasks such as the following [3,24]:

- Works with market researchers
- Helps to develop and implement appropriate questionnaires
- Performs analysis of industry and regulatory standards
- Evaluates competitive devices
- Conducts interviews with potential users of the device

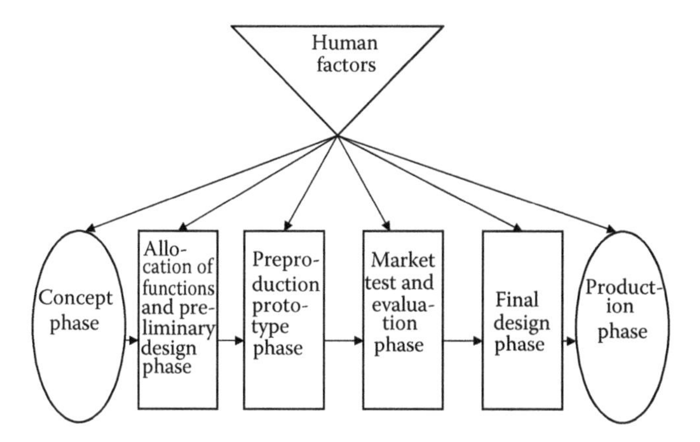

FIGURE 9.2
Medical device development process with human factors inputs.

- Examines the proposed operation of the potential device with respect to skill range, educational background, and experiences of the intended users
- Identifies all the possible use environments of the device under consideration

During the allocation of functions and preliminary design phase, the human factors specialist works alongside the design professionals to determine which device functions will require manual points of interface between humans and the device, and which will be automatic. In regard to the points of interface, they are operations where humans are required to monitor and control, so that the desired output or feedback from the device is obtained. Thus, the preliminary design analysis is performed with respect to the operational environment of the device under consideration and the skill levels of the most untrained device users. Normally, this task is conducted by carefully considering the sketches/drawings of the expected operational device environment and gauging reactions of all possible users.

During the preproduction prototype phase of the device, the prototype is constructed or updated for additional evaluation and market-related testing. The market test and evaluation phase involves, in addition to the actual testing of the medical device, a thorough examination of the feedback received from the market test by human factors and other professionals.

During the final design phase, the design of the medical device is finalized by incorporating human-factors-related changes generated by the test, evaluation, and marketing. Finally, in the production phase, the medical device is manufactured and put on the market. During this phase the human factors professional normally monitors the performance of the device, performs analysis of the proposed changes to the device design, and provides appropriate assistance in the development of user training programs.

9.9 Problems

1. Write an essay on medical device safety and errors.
2. Discuss the following two types of medical device safety:
   - Unconditional safety
   - Descriptive safety
3. What are the three main areas of requirements placed by the government and other agencies on medical devices, directly or indirectly, with respect to safety?
4. Discuss the legal aspects of medical device safety.
5. Discuss software-related issues in the safety of cardiac rhythm management products.
6. List seven main classifications of medical device accidents.
7. What are the 10 most error-prone medical devices?
8. Discuss medical device operator errors.
9. Describe the general approach to human factors during the medical device development process for reducing human errors.
10. Write an essay on medical device electrical safety standards.

9.10 References

1. Goetsch, D. L., *Occupational Safety and Health*, Prentice-Hall, Englewood Cliffs, New Jersey, 1996.
2. LaDou, J., Editor, *Introduction to Occupational Health and Safety*, National Safety Council, Chicago, Illinois, 1986.
3. Dhillon, B. S., *Medical Device Reliability and Associated Areas*, CRC Press, Boca Raton, Florida, 2000.
4. Dhillon, B. S., Reliability Technology in Health Care Systems, *Proceedings of the IASTED International Symposium on Computers and Advanced Technology in Medicine, Health Care, and Bio-engineering*, 1990, pp. 84–87.
5. Maddox, M. E., Designing Medical Devices to Minimize Human Error, *Medical Device & Diagnostic Industry Magazine*, Vol. 19, No. 5, 1997, pp. 166–180.
6. Nobel, J. L., Medical Device Failures and Adverse Effects, *Pediatric Emergency Care*, Vol. 7, 1991, pp. 120–123.
7. Bogner, M. S., Medical Devices and Human Error, in *Human Performance in Automated Systems: Current Research and Trends*, edited by M. Mouloua, R. Parasuraman, Lawrence Erlbaum Associates, Hillsdale, New Jersey, 1994, pp. 64–67.
8. Leitgeb, N., *Safety in Electromedical Technology*, Interpharm Press, Buffalo Grove, Illinois, 1996.

9. Fries, R. C., *Reliable Design of Medical Devices*, Marcel Dekker, New York, 1997.
10. Lyn, M. R., Editor, *Handbook of Software Reliability Engineering*, McGraw-Hill, New York, 1996.
11. Pecht, M., Editor, *Product Reliability, Maintainability, and Supportability Handbook*, CRC Press, Boca Raton, Florida, 1995.
12. Dhillon, B. S., *Design Reliability: Fundamentals and Applications*, CRC Press, Boca Raton, Florida, 1999.
13. Bethune, J., Editor, On Product Liability: Stupidity and Waste Abounding, *Medical Device & Diagnostic Industry Magazine*, Vol. 18, No. 8, 1996, pp. 8–11.
14. Marcus, M. L., Biersach, B. R., Regulatory Requirements for Medical Equipment, *IEEE Instrumentation and Measurement Magazine*, December 2003, pp. 23–29.
15. UL2601-1, *Medical Electrical Equipment—Part 1, General Requirements for Safety*, Underwriters Laboratories, Northbrook, Illinois, 1997.
16. IEC 606D1-1, *Medical Electrical Equipment Part 1, General Requirements for Safety*, International Electrotechnical Commission (IEC), Geneva, Switzerland, 1988.
17. Marcus, M. L., Biersach, B. R., Avoiding Last Minute Redesign of Medical Electrical Equipment—Current Certification and Product Safety Requirements, *Medical Device & Diagnostic Industry Magazine*, Vol. 25, No. 3, 2003, pp. 90–99.
18. Mojdehbakhsh, R., et al., Retrofitting Software Safety in an Implantable Medical Device, *IEEE Software*, Vol. 11, January 1994, pp. 41–50.
19. Lowen, B., Cardiovascular Collapse and Sudden Cardiac Death, in *Heart Disease Textbook of Cardiovascular Medicine*, W. B. Saunders, Philadelphia, 1984, pp. 778–808.
20. Brueley, M. E., Ergonomics and Error: Who Is Responsible?, *Proceedings of the First Symposium on Human Factors in Medical Devices*, 1989, pp. 6–10.
21. Hammer, W., *Product Safety Management and Engineering*, Prentice Hall, Englewood Cliffs, New Jersey, 1980.
22. Wikland, M. E., *Medical Device and Equipment Design*, Interpharm Press, Buffalo Grove, Illinois, 1995.
23. Hyman, W. A., Human Factors in Medical Devices, in *Encyclopaedia of Medical Devices and Instrumentation*, edited by J. G. Webster, Vol. 3, John Wiley and Sons, New York, 1988, pp. 1542–1553.
24. Le Cocq, A. D., Application of Human Factors Engineering in Medical Product Design, *Journal of Clinical Engineering*, Vol. 12, No. 4, 1987, pp. 271–277.

10

Medical Device Usability

10.1 Introduction

Each year, a vast sum of money is spent to produce various types of medical devices throughout the world. Their usability has become an important issue, because various studies conducted over the years indicate that poorly designed human–machine interfaces of medical devices significantly increase the risk for the occurrence of human errors [1–4]. These errors can directly or indirectly result in patient injury or death.

Medical device usability may be defined as the medical device's interactive systems quality with respect to factors such as user satisfaction, ease of use, and ease of learning [1,5]. This means that to attain high user adoption and successfully navigate all involved regulatory processes, a medical device must be designed around all types of users and must incorporate proper defenses against potential risks under varied conditions. More specifically, the medical device usability-related concerns with respect to users such as patients, nurses, professional caregivers, physicians, and family members must be raised to the same level as conventional economic, manufacturing, and technological concerns, during the design phase.

This chapter presents various important aspects of medical device usability.

10.2 Medical Device Users and Use Environments

There are various types of medical device users who need devices that can be used safely and effectively. To satisfy these user needs, it is important to clearly understand the limitations and abilities of all medical device potential users, i.e., patients, professional health care providers, and young and old individuals. Factors that can affect the ability of such users include fatigue, stress, and medication. Some of the important characteristics of the medical device potential users that must be taken into consideration during the device design process are shown in Figure 10.1 [6].

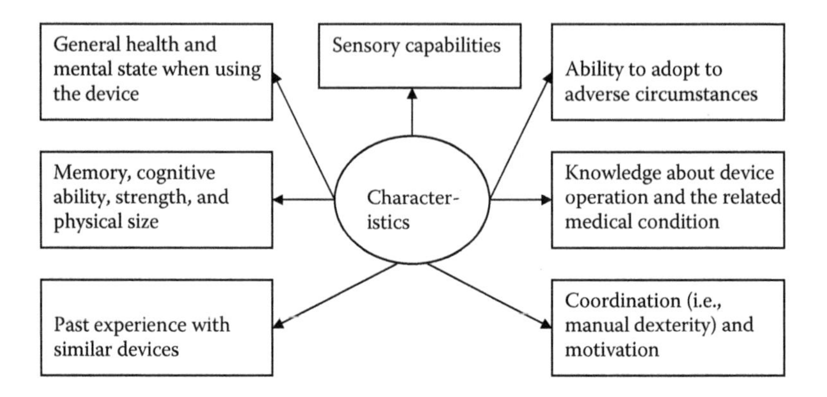

FIGURE 10.1
Important characteristics of the medical device potential users for consideration during the device design process.

The use environments of medical devices can vary significantly from one situation to another and can impact their usability significantly. The main factors with respect to device users that must be considered with care are as follows [2]:

- Noise and light
- Vibration and motion
- Mental workload
- Physical workload

In the case of *noise and light*, the effectiveness of auditory and visual displays (e.g., auditory alarms and light indicators) can be limited by the use environments if they are not designed properly. For example, in noisy environments, alarms may not be heard by the device users if they are not distinctive or sufficiently loud. In the case of *vibration and motion*, these two things can seriously affect device users' ability to read displayed information, to carry out fine physical manipulations such as typing on the keyboard part of a medical device, and so on.

Mental workload is concerned with the degree of thinking and concentration a person exerts while using a medical device. When the mental workload imposed on device users by the environments exceeds their abilities, the chances of error occurrence in using medical devices increase dramatically. An example of such situations could be an operating room having too many alarms on different medical devices, thus making it very difficult for anesthetists to accurately identify the source of any single alarm.

Finally, *physical workload* is concerned with the use of the medical device and adds to user stress. Under high stress, the users of medical devices are distracted and have less time to make decisions (e.g., consider multiple device outputs).

10.3 Medical Device User Interfaces and Use Description

The user interfaces of medical devices are important because they help to facilitate correct actions and prevent or discourage the occurrence of hazardous actions. They comprise all elements of medical devices with which users interact while using devices, preparing them for use (e.g., setup and calibration), or performing maintenance activities. The user interface incorporates all hardware features that control the device operation. Some examples of these features are as follows [2,6]:

- Knobs
- Buttons
- Switches
- User information providing device features such as indicator lights, displays, and visual and auditory alarms

In medical devices, user interfaces are normally computer-based. Thus, in this case interface characteristics include items such as keyboards, mouse, control and monitoring screens, data entry requirements, prompts, the manner in which data are organized and presented, alerting mechanisms, help functions, navigation logic, and screen elements [2,6]. Items such as device labeling, operating instructions, packaging, and training materials are also considered part of the user interface.

The clearly written use description of a medical device is essential for accurate and complete understanding of its use. The description includes information on items such as device operation, user interface design or preliminary design, use environments, user population characteristics, general use scenarios, and user needs for effective and safe use of the device and how the device satisfies them [2,6].

10.4 An Approach to Develop Medical Devices' Effective User Interfaces

A six-step general approach to develop effective user interfaces for medical devices is as follows [1,2,7]:

- **Step 1:** Define project goals and system functionality.
- **Step 2:** Conduct analysis of user tasks, work environments, and capabilities and allocate tasks between the humans and the system.

- **Step 3:** Document user needs and requirements, and establish usability goals and design prototypes.
- **Step 4:** Conduct usability testing and evaluate its results against performance goals and objectives and make a loop back to Step 3 as appropriate.
- **Step 5:** Develop design specifications for device user interface.
- **Step 6:** Evaluate device interface designs during their field use and make loop backs to Step 3 as considered appropriate.

Additional information on this approach is available in Refs. [1,7].

10.5 Useful Guidelines to Reduce Medical Device/Equipment User-Interface-Related Errors

Medical devices such as ventilators, blood chemistry analyzers, patient monitors, infusion pumps, and kidney dialysis machines often have various types of user-interface design-related problems. These problems can negatively affect usability and appeal of a medical device. The following guidelines address these user-interface design problems, in turn reducing device user-interface-related errors [8].

- **Minimize design inconsistencies.** These inconsistencies are harmful to usability and user-interface appeal, and for some medical devices, they can compromise safety.
- **Minimize screen density.** This is concerned with lowering the overstuffing of displays of medical devices with information and controls. The resulting empty space by such reduction is useful in a user interface because it helps to separate information into related categories/groups and provides a resting place for users' eyes. Overly dense device user interfaces can be intimidating to nurses, technicians, and physicians, making it difficult for these professionals to retrieve required information at a glance. Actions such as presented in Table 10.1 can be useful to eliminate extraneous information on the displays of medical devices.
- **Maximize simplification of typography.** User interfaces of efficient medical devices are based on typographical rules that make screen contents easy to read and direct users effectively toward the important information first. This can be achieved by having a few character sizes and a single font. Another mechanism used for simplifying typography is to minimize excessive highlighting such as underlining, bolding, and italicizing.

TABLE 10.1

Useful Actions to Eliminate Extraneous Information on Medical Device Displays

No.	Action
1	Use empty space rather than lines to separate content
2	Use simplified graphics
3	Decrease text size by stating things in a more simplified manner
4	Reduce the size of graphics concerned with identity (i.e., brand names and logos)
5	Present secondary information on demand through pop-ups or relocate it to other screens (if possible)

- **Limit the usage of colors.** This is concerned with limiting the color palette of medical devices' user interfaces. The following two guidelines can be used:
    - Ensure that the selection of colors is consistent with existing medical conventions. For example, red is frequently used to communicate arterial blood pressure values or to symbolize alarm-related information.
    - Limit the number of colors of the background and major on-screen elements to between three and five, including shades of gray.
- **Maximize the use of hierarchical labels.** As the use of redundant labels leads to congested screens that can take a long time to scan, hierarchical labeling is useful to save space and speed scanning by displaying items such as heart rate, respiratory rate, and arterial blood pressure more efficiently.
- **Aim to ascribe to a grid.** Screens operate and look better when their components are aligned and serve a utilitarian objective effectively. Grid-based screens are usually easier to implement in computer code because of visual elements' predictability. Two guidelines with respect to ascribing to a grid are as follows:
    - Keep on-screen elements or components at a fixed distance from the gridlines.
    - Begin by defining the dominant elements or components of the screen and approximate space requirements when developing a grid structure.
- **Maximize the usage of simple language.** Medical device user interfaces are often characterized by overly complex phrases and words that lead to usability problems. Some corrective measures are giving meaningful subheadings and headings, writing shorter sentences, breaking difficult procedures into a number of ordered steps, and using consistent syntax.
- **Create effective visual balance.** Normally, this is created about the vertical axis by arranging visual elements on either side of an

assumed axis with each side containing about the same amount of content as empty space. There are a number of approaches that can be used to evaluate the balance of a composition, and perceived imbalances can be corrected through means such as reorganizing information, popping up elements only upon request, and relocating elements to other screens.

- **Refine and harmonize icons.** Some useful actions that can give the icons a family resemblance to each other and to maximize icon comprehension are as follows [2,8]:
  - Simplify icon elements to eliminate confusing and unnecessary details.
  - Use the same style for all similar-purpose icons.
  - Develop a limited set of icon elements that represent nouns only.
  - Perform user testing to ensure that no two icons are so similar they create confusion.
  - Make similar-purpose icons the same size.
  - Reinforce all icons with text labels.
- **Provide appropriate navigation options and cues.** In a medical device user interface, a navigator can sometimes become lost when going from one place to another. Actions such as those shown in Figure 10.2 show some appropriate navigation options and cues.

10.6 Designing and Developing Medical Devices for Older User Population

The population of older people in the United States is growing at a significant rate, and it is forecast that by the year 2020 one out of every five or six Americans will be over 65 years of age [2,9,20]. This means there is a definite need to consider the following factors during the medical device design [2,9,11–13]:

- Physical limitations
- Sensory limitations
- Cognitive limitations

In regard to physical limitations, a significant number of people usually lose 10% to 20% of their strength by the time they reach 60 to 70 years of age. Their mobility may also be limited by various types of joint-related problems. To overcome physical limitations such as these, designers of medical

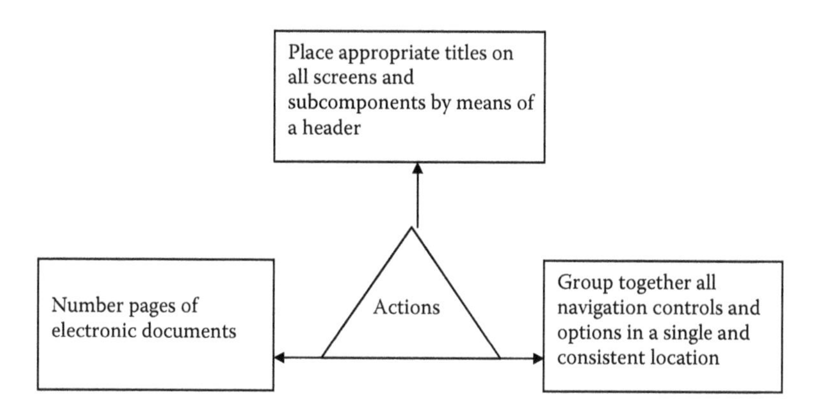

FIGURE 10.2
Useful actions for providing appropriate navigation options and cues.

devices should incorporate controls with large-diameter knobs so that rotation requires lesser fine control, textured knob surfaces that require less pinching strength to overcome finger slippage, and so on.

In regard to sensory limitations, the two common limitations among older people are impaired vision and hearing. To overcome impaired vision limitation, the designers must give careful consideration to using larger fonts for readouts, displays, and labels on medical devices. Decline in the hearing ability of an individual is usually a function of age. Females and males, as they age, suffer greater hearing loss at frequencies in the 550–1000 Hz range and 3000–6000 Hz range, respectively. Designers must consider these factors in medical devices to be used by older individuals.

Finally, in regard to cognitive limitations, cognition abilities of older people can vary quite significantly. Some may experience attention deficits frequently referred to as *cognitive rigidity*, a condition that makes it difficult to learn new procedures and approaches, so designers must reduce the number of steps in a given procedure concerning a medical device to improve its usability among the older population [2,9].

A general recommendation to effectively facilitate user tasks for the older population is building redundant cues in the design of medical devices [2,14].

10.7 Guidelines for Designing Hand-Operated Devices with Respect to Cumulative Trauma Disorder

Cumulative trauma disorder is most prevalent among people whose occupations demand the performance of the same task repeatedly. The affliction rate for cumulative trauma disorders is as high as 25% among individuals performing motion-intensive tasks with their hands [15].

Cumulative trauma disorder presents a significant risk to health care workers and can be minimized by better medical device usability-related design and better work habits reinforced through effective instructions and warning labels [2,16].

Some guidelines to design hand-operated medical devices with respect to cumulative trauma disorder are presented below [2,9,17]:

- Conduct analysis of the range of user hand motion as a basis to determine the dynamic characteristics of handles and controls.
- Shield devices to reduce vibrations they will transmit to their potential users.
- Ensure that gripping controls and surfaces are designed in such a way that they enable all types of potential users to keep their hands in a neutral, resting position.
- Provide effective advisory instructions/visual cues in regard to holding a device.
- Reduce the weight of objects so that they can easily be moved or picked up.
- Provide appropriate padding and ergonomically contoured surfaces to reduce concentration of mechanical stresses on the user's skin and underlying tissues.
- Design hand-operated devices in such a way that they will be comfortable for persons with different hand sizes.
- Avoid designs that will require device users to exert force all the time.
- Provide appropriate force-assist mechanisms as necessary to reduce the muscle exertion required to operate a device.
- Develop appropriate operational sequences that stop the frequent occurrence of a repetitive movement.
- Provide effective instructions to device users on how to prevent the occurrence of cumulative trauma disorders.
- Position work surfaces in such a manner that permits forearms to extend at an angle of about 90 degrees with respect to the user's body, with the elbows held at one's side.
- Ensure that an adequate amount of space is provided for hand and forearm movements to stop device users from assuming poor hand postures while performing tasks.
- Select materials for handles that provide a nonslip grip and protection to device users' hands from cold temperatures, electrical conduction, and vibrations.
- Design objects so that they can easily be grasped by the entire hand, rather than pinched between the thumb and fingers, in situations when high precision is not required.

- Design awkwardly shaped or heavy objects in such a way that they can easily be lifted or grasped by device users with both hands.

10.8 Useful Documents for Improving Usability of Medical Devices

Over the years, a large number of publications on engineering usability and related areas have appeared [2]. Some that can be useful to improve medical device usability—and in turn, patient safety—are as follows:

- Obradovich, J. H., Woods, D. D., Users as Designers: How People Cope with Poor HCI Design in Computer-Based Medical Devices, *Human Factors*, Vol. 38, No. 4, 1996, pp. 574–592.
- Liljegren, E., Osvalder, A. L., Cognitive Engineering Methods as Usability Evaluation Tools for Medical Equipment, *International Journal of Industrial Ergonomics*, Vol. 34, No. 1, 2004, pp. 49–62.
- Clans, P. L., Gibbons, P. S., Kaihoi, B. H., Usability Laboratory: A New Tool for Process Analysis at the Mayo Clinic, *Proceedings of the Healthcare Information Management Systems Society Conference*, 1997, pp. 149–159.
- Brown, S., The Challenges of User-Based Design in a Medical Equipment Market, in *Field Methods Casebook for Software Design*, John Wiley and Sons, New York, 1996, pp. 157–176.
- Gosbee, J., The Discovery Phase of Medical Device Design: A Blend of Intuition, Creativity, and Science, *Medical Device and Diagnostic Industry Magazine*, October 1997, pp. 113–118.
- Hyman, W. A., Errors in the Use of Medical Equipment, in *Human Error in Medicine*, Lawrence Erlbaum Associates, Hillsdale, New Jersey, 1994, pp. 327–347.
- Jan, A., Lauler, W., Radermacher, K., *IFMBE Proceedings*, Vol. 25, No. 12, 2009, pp. 180–183.
- Garmer, K., Lijegren, E., Osvalder, A. L., Dahlman, S., Application of Usability Testing to the Development of Medical Equipment: Usability Testing of a Frequently Used Infusion Pump and a New User Interface Developed with a Human Factors Approach, *International Journal of Industrial Ergonomics*, Vol. 29, No. 3, 2002, pp. 145–159.
- Cook, R. I., Woods, D. D., Adapting to New Technology in the Operating Room, *Human Factors*, Vol. 38, No. 4, 1996, pp. 593–613.
- Gosbee, J., Ritchie, E. M., Human-Computer Interaction and Medical Software Development, *Interactions*, Vol. 4, No. 4, 1997, pp. 13–18.

- Navai, M., Guo, X., Caird, J. K., Dewar, R. E., Understanding of Prescription Medical Labels as a Function of Age, Culture, and Language, *Proceedings of the Human Factors and Ergonomics Society Conference*, 2000, pp. 1487–1491.
- Morrow, D. G., Leirer, V. O., Andrassy, J. M., Using Icons to Convey Medication Schedule Information, *Applied Ergonomics*, Vol. 27, 1996, pp. 267–275.
- Seagull, F. J., Sanderson, P. M., Anesthesia Alarms in Context: An Observational Study, *Human Factors*, Vol. 43, 2001, pp. 66–78.
- Wiklund, M., Editor, *Usability in Practice: How Companies Develop User-Friendly Products*, Academic Press, Cambridge, Massachusetts, 1994.
- *Designer's Handbook: Medical Electronics*, Canon Communications, Santa Monica, California, 1995.
- Chaffin, D. B., Faraway, J. J., Zhang, X., Woolley, C., Stature, Age, and Gender Effects on Reach Motion Postures, *Human Factors*, Vol. 42, 2000, pp. 408–420.
- Hix, D., Hartson, R., *Developing User Interfaces: Ensuring Usability through Product and Process*, John Wiley and Sons, New York, 1993.
- Burgess-Limerick, R., Mon-Williams, M., Coppard, V. L., Visual Display Height, *Human Factors*, Vol. 42, 2000, pp. 140–150.
- ANSI/AAMI-HE-48, *Human Factors Engineering Guidelines and Preferred Practices for the Design of Medical Devices*, American National Standards Institute (ANSI), New York, 1993. This standard was developed by the Association for the Advancement of Medical Instrumentation (AAMI) and approved by the ANSI, AAMI, Arlington, Virginia, 1993.

10.9 Problems

1. Write an essay on medical device usability.
2. What are the important characteristics of the medical device potential users to consider during the design process?
3. Discuss medical device user interfaces.
4. Describe the approach for developing effective user interfaces for medical devices.
5. What are the useful actions to eliminate extraneous information on medical device displays?

6. What are the important factors that must be considered during the medical device design when the device is going to be used by individuals over 65 years of age?
7. List at least 10 useful guidelines for designing hand-operated devices with respect to cumulative trauma disorder.
8. List seven most useful documents to improve usability of medical devices.
9. Discuss at least five useful guidelines to reduce medical device/equipment user-interface-related errors.
10. Write an essay on medical device use environments.

10.10 References

1. Garmer, K., et al., Arguing for the Need of Triangulation and Iteration When Designing Medical Equipment, *Journal of Clinical Monitoring and Computing*, Vol. 17, 2002, pp. 105–114.
2. Dhillon, B. S., *Engineering Usability: Fundamentals, Applications, Human Factors, and Human Error*, American Scientific Publishers, Stevenson Ranch, California, 2004.
3. Hayman, W. A., Errors in the Use of Medical Equipment, in *Human Error in Medicine*, edited by M. S. Bogner, Lawrence Erlbaum Associates, New York, 1994, pp. 327–347.
4. Obradovich, J. H., Woods, D. D., Users as Designers: How People Cope with Poor HCI Design in Computer-Based Medical Devices, *Human Factors*, Vol. 38, 1996, pp. 40–46.
5. ISO 13407, *User-Centered Design Process for Interactive Systems*, International Organization for Standardization (ISO), Geneva, Switzerland, 1999.
6. *Medical Device Use-Safety: Incorporating Human Factors Engineering into Risk Management*, Draft Guidance Document, Center for Devices and Radiological Health, Food and Drug Administration, Washington, D.C., 2000.
7. Salvemini, A. J., Challenge for User-Interface Designers of Telemedicine Systems, *Telemedicine Journal*, Vol. 5, No. 2, 1999, pp. 10–15.
8. Wiklund, M. E., Making Medical Device Interfaces More User Friendly, *Medical Device and Diagnostic Industry (MDDI) Magazine*, May 1998, pp. 177–184.
9. Wiklund, M. E., *Medical Device and Equipment Design: Usability Engineering and Ergonomics*, Interpharm Press, Buffalo Grove, Illinois, 1995.
10. Czaja, S., Special Issue Preface, *Human Factors*, Vol. 32, No. 5, 1990, pp. 505.
11. CPSC Publication No. 702, *Product Safety and the Older Consumer: What Manufacturers/Designers Need to Consider*, Consumer Product Safety Commission (CPSC), Washington, D.C., 1988.
12. Small, A., Design for Older People, in *Handbook of Human Factors*, edited by G. Salvendy, John Wiley and Sons, New York, 1987, pp. 125–140.
13. Czaja, S., Clark, M., Weber, R., Computer Communication among Older Adults, *Proceedings of the Human Factors Society 34th Annual Meeting*, 1990, pp. 304–309.

14. Koncelik, J., *Aging and the Product Environment*, Scientific and Academic Additions, Florence, Kentucky, 1982.
15. Armstrong, T., Radwin, R. G., Hansen, D. J., Repetitive Trauma Disorders: Job Evaluation and Design, *Human Factors*, Vol. 28, No. 3, 1986, pp. 325–330.
16. Herbert, L., *Living with CTD*, IMPCC, Bangor, Maine, 1990.
17. Putz-Anderson, V., *Cumulative Trauma Disorders*, Taylor and Francis, New York, 1988.

11

Patient Safety Organizations, Data Sources, and Mathematical Models

11.1 Introduction

In recent years patient safety has become an important issue because of a staggering number of deaths and injuries due to patient safety-related problems. For example, as per an Institute of Medicine report around 100,000 Americans die each year due to human errors in the health care system [1]. Today there are many patient safety organizations in various parts of the world that advocate improvement in patient safety. A patient safety organization may be described as a group, association, or institution that improves medical care by reducing the occurrence of medical errors.

Patient safety-related data plays a pivotal role in decisions concerning patient safety in the health care system. The effectiveness of such decisions depends on the quality of patient safety data (i.e., poor quality data will lead to ineffective or poor decisions). This means that careful attention is required when collecting and analyzing such data. Nonetheless, there are many good sources for collecting various types of patient safety-related data.

Mathematical models are often used in engineering to study various types of physical phenomena. Over the years, a large number of mathematical models have been developed to study human reliability and error in engineering systems [2]. Some of these models can also be used to study patient safety-related problems.

This chapter presents various important aspects of patient safety organizations, data sources, and mathematical models useful for performing patient safety analysis.

11.2 Patient Safety Organization Types and Functions

Over the years, various types of patient safety organizations have been established ranging from governmental to private, and some founded by industry, consumer, or professional groups. They can be grouped under two categories: governmental organizations and independent organizations. Some examples of the governmental organizations are the Agency for Healthcare

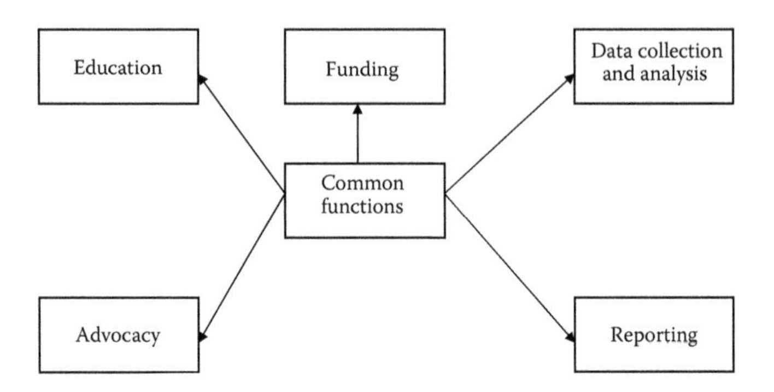

FIGURE 11.1
Common functions of patient organizations.

Research and Quality in the United States, the National Patient Safety Agency in the United Kingdom, and the World Alliance for Patient Safety launched by the World Health Organization (WHO) in October 2004. Similarly, three examples of the independent organizations are the National Patient Safety Foundation in the United States, the Canadian Patient Safety Institute in Canada, and the Australian Patient Safety Foundation in Australia.

The common functions of patient safety organizations are shown in Figure 11.1. More specifically, to reduce the occurrence of adverse events these organizations use approaches such as follows:

- Conduct fundraising and provide funding for patient safety-related research projects.
- Collect data on the prevalence and individual details of patient safety-related errors.
- Analyze error sources by using the root cause analysis (RCA) method [3].
- Advocate for necessary legislative and regulatory changes.
- Propose and disseminate appropriate error prevention methods.
- Raise awareness and inform the health care professionals, providers, public, employers, and purchasers.
- Design and perform pilot projects to study patient safety-related initiatives, including monitoring of end results.

11.3 Governmental Patient Safety Organizations

Many patient safety organizations around the world have been established that look after and promote patient safety from the government perspective. Some of these organizations are described below.

11.3.1 World Alliance for Patient Safety (WAPS)

This patient safety organization was launched by WHO in October 2004, with its goal to develop standards for patient safety and provide appropriate assistance to United Nations (UN) member states to improve health care safety [4]. The WAPS raises awareness and political commitment to improve safety in addition to facilitating the development of patient safety policies and practices in all WHO member states. Each year, WAPS delivers a number of programs around the globe that cover systemic and technical aspects to improve patient safety.

Until May 2006, WAPS held patient safety-related meetings in five of the six WHO regions, in addition to about 40 technical workshops concerning patient safety in 18 countries [4]. Ever since its launch in October 2004, WAPS has made significant progress in areas such as WHO guidelines on Adverse Event Reporting and Learning Systems, development of a patient safety taxonomy to classify data on patient safety problems, and performance of prevalence study on patient harm in developing countries [4].

11.3.2 Agency for Healthcare Research and Quality (AHRQ)

This agency is a part of the U.S. Department of Health and Human Services. The main mission of AHRQ is to improve the safety, quality, and effectiveness of health care for American people. It organizes patient safety activities, serves as a clearinghouse for safety information, provides grants to other organizations, publishes guidelines for evidence-based or best practices, and so on [5].

In partnership with data organizations in 37 states, AHRQ sponsors the Nationwide Inpatient Sample (NIS), a data bank of the Healthcare Cost and Utilization Project (HCUP). The HCUP is a federal-state-industry partnership that provides all discharge-related data from over 990 hospitals (i.e., about 8 million annual hospital stays) [5]. The NIS is the largest all-payer inpatient care data bank in the United States, and it can be used to derive national estimates of inpatient care. By using the data from this data bank, the AHRQ has been able to provide complication rates and risk-related data, even for quite rare surgical procedures, such as bariatric surgery [6].

11.3.3 National Patient Safety Agency (NPSA)

This agency was created in July 2001 in the United Kingdom to improve patient safety within the framework of the UK National Health Service (NHS) by encouraging voluntary reporting of medical-related errors, performing analysis, and initiating appropriate preventive measures. Since 2005, the NPSA has also been responsible for the following items [7]:

- Safety aspects of hospital design, cleanliness, and food
- Safe research practices through the UK National Research Ethics Service (NRES)
- Performance of individual doctors and dentists, through the UK National Clinical Assessment Service (NCAS)

The NPSA identifies patient safety-associated deficiencies with the aid of input from patients and clinical experts, develops appropriate solutions, and monitors results of corrective measures within the NHS. Its initiatives and alerts include items such as hand hygiene, information for doctors and patients on steps to decrease risk of error, vaccine safety, and disclosure of error to all injured patients. Finally, the National Reporting and Learning System (NRLS) allows all NHS employees to provide the NPSA with reports on an anonymous basis.

11.3.4 Australian Council for Safety and Quality in Health Care (ACSQHC)

This council was established by the Australian government in January 2000 to provide leadership in improving patient safety and quality through advice to all federal, state, and territory health ministers for a five-year term. Its three main specific goals were as follows:

- To develop common standards for patient safety
- To improve data collection for safety monitoring
- To provide for public input at the governmental level

In January 2006, the council was replaced by the Australian Commission on Safety and Quality in Health Care with the same role. The annual reports of the ACSQHC included recommendations such as developing a national strategy for preventing health care-associated infections, a national trial of disclosure of adverse events, and national credentialing for medical practitioners in hospitals.

11.4 Independent Patient Safety Organizations

There are probably more independent patient safety organizations today than there are governmental patient safety organizations around the world. Some of these organizations are described below.

11.4.1 National Quality Forum (NQF)

This is a not-for-profit membership organization established in 1999 in the United States, for developing and implementing a national strategy for health care quality measurement and reporting [8]. Its membership is open to national, state, regional, and local groups representing employers, consumers, private and public purchasers, health care professionals, provider organizations, accrediting bodies, labor unions, supporting industries, and organizations involved in the area of health care quality improvement or research. Over the years, NQF has focused on several areas including error rates, unnecessary procedures and undertreatments, and preventive care.

In 2002, NQF defined 27 events that must never occur within the framework of a health care facility. They include surgical events (e.g., surgery performed on the wrong patient), care management events (e.g., a medication error), patient protect events (e.g., an infant discharged to the wrong person), environmental events (e.g., burn or electric shock), criminal events (e.g., sexual assault of a patient), and product or device events (e.g., using contaminated drugs). The organization has many ongoing projects and it recommends a national state-based event reporting system for improving the quality of patient care.

11.4.2 Australian Patient Safety Foundation (APSF)

This organization was founded in 1989 to monitor anesthesia errors and was expanded to patient incident reporting and monitoring after the findings of the Quality in Australian Health Care Study (QAHCS) in 1995 [9]. Adverse medical events are reported and analyzed through its subsidiary known as Patient Safety International (PSI), using the Advanced Incident Management System (AIMS), a software tool.

AIMS is used in over 50% of the hospitals in Australia and in 2005 it was adopted by the New Zealand Accident Compensation Corporation. The AIMS data remains confidential and is protected from legal actions under the Australian Commonwealth Quality Assurance legislation.

11.4.3 National Patient Safety Foundation (NPSF)

This organization was jointly founded by the American Medical Association along with two other organizations in 1996 and is based on the model of the Anesthesia Patient Safety Foundation. The NPSF provides leadership, research support, and education in regard to patient safety in the United States. Since 1998, NPSF has been holding an Annual Patient Safety Congress to promote patient safety-related research in the country.

A scientific journal entitled *Journal of Patient Safety* is published by the NPSF. The journal regularly publishes original and review articles on various aspects of patient safety.

11.4.4 Canadian Patient Safety Institute (CPSI)

The CPSI was established in 2003 after inputs from various Canadian health care professional organizations and federal, provincial, and territorial ministries of health [10]. It is an independent organization and promotes solutions and collaboration among governments and stakeholders to improve patient safety. The main areas of improvement are education, communication, system innovation, regulatory affairs, and research.

In 2005, CPSI launched the "Safer Healthcare Now" campaign to reduce medical error-related injuries by focusing on a number of evidence-based measures.

11.4.5 Joint Commission on Accreditation of Healthcare Organizations (JCAHO)

The JCAHO was founded in 1951 and is an independent organization that evaluates and accredits around 15,000 health care organizations and programs in the United States [11]. The scope of reviews by JCAHO is quite broad; it includes hospitals, nursing homes, surgical centers, home care agencies, medical laboratories, medical equipment providers, and so on. Since 1966, identifying sentinel events and analyzing the root causes has been a main focus of JCAHO. Its first eight alerts were published in 1998.

In 2005, JCAHO founded an International Center for Patient Safety to collaborate with international patient safety organizations for identifying, developing, and sharing safety-related solutions, advocating public policy changes, and performing joint research.

11.5 Data Sources

Patient safety-related data are the backbone of patient safety-related studies. They are the final proof of the efforts expended to improve patient safety in health care facilities, medical device/equipment design, and so on. There are various sources of obtaining patient safety-related data. Some organizations that offer patient safety-related data are as follows [3]:

- National Patient Safety Agency, 4-8 Maple Street, London W1T 5HD, United Kingdom
- National Patient Safety Foundation, 132 Mass Moca Way, North Adams, MA 01247
- World Alliance for Patient Safety, c/o World Health Organization, 20 Avenue Appia CH-1211, Geneva 27, Switzerland

- Canadian Patient Safety Institute, Suite 1414, 10235-101 Street, Edmonton, Alberta, Canada
- Safety and Quality in Health Care, Health Care Division, Department of Health, P.O. Box 8172, Stirling Street, Perth, Australia
- Agency for Healthcare Research and Quality, 500 Gaither Road, Rockville, MD 20857-0001
- Australian Patient Safety Foundation, P.O. Box 400, Adelaide 5001, Australia
- National Quality Forum, 601 13th Street NW, Suite 500 North, Washington, D.C.
- The Joint Commission on Accreditation of Healthcare Organizations (JCAHO), 1 Renaissance Blvd., Oakbrook Terrace, Illinois
- U.S. Food and Drug Administration, 5600 Fishers Lane, Rockville, MD 20857-0001
- Institute for Safe Medication Practices, 1800 Byberry Road, Suite 810, Huntingdon Valley, PA 19006

11.6 Mathematical Models

Various types of mathematical models have been developed to perform human performance reliability analysis in the area of engineering. Some of these models can also be used in the area of patient safety to conduct human performance reliability analysis. Four of these models are presented below.

11.6.1 Model I

This model is concerned with predicting the reliability of health care professionals performing various types of time-continuous tasks including operating, monitoring, and tracking. The reliability of health care professionals performing tasks such as these can be estimated by using the following equation [2,3]:

$$RH_{cp}(t) = e^{-\int_0^t \lambda_{cp}(t)dt} \tag{11.1}$$

where

$RH_{cp}(t)$ = the reliability of the health care professional at time t.

$\lambda_{cp}(t)$ = the time-dependent error rate of the health care professional.

The derivation of Equation (11.1) is given in Ref. [2]. The equation can be used to predict the reliability of a health care professional when his/her time to error follows any time-continuous probability distribution (e.g., exponential, normal, or Weibull).

Example 11.1

Assume that the error rate of a health care professional is 0.008 errors per hour. Calculate his/her reliability for a 10-hour mission.

By substituting the specified data values into Equation (11.1), we get

$$RH_{cp}(10) = e^{-\int_0^{10} (0.008)dt}$$

$$= e^{-(0.008)(10)}$$

$$= 0.9231$$

Thus the health care professional's reliability is 0.9231.

11.6.2 Model II

This mathematical model is concerned with predicting the mean time to error of health care professionals performing time-continuous tasks such as operating, monitoring, and tracking. This can be estimated by using the following equation [2]:

$$MTTEH_{cp} = \int_0^{\infty} \left[e^{-\int_0^t \lambda_{cp}(t)dt} \right] dt \tag{11.2}$$

where

$MTTEH_{cp}$ = the health care professional's mean time to error.

Equation (11.2) can be used to predict mean time to error of a health care professional when his/her time to error follows any time-continuous probability distribution (e.g., exponential, Weibull, or normal).

Example 11.2

Assume that the error rate of a health care professional performing his/her assigned tasks is 0.005 errors per hour. Calculate his/her mean time to error.

By substituting the specified data value into Equation (11.2), we obtain

$$MTTEH_{cp} = \int_0^{\infty} \left[e^{-\int_0^t (0.005)dt} \right] dt$$

$$= \frac{1}{0.005}$$

$$= 200 \text{ hours}$$

Thus the health care professional's mean time to error is 200 hours.

11.6.3 Model III

This model is concerned with predicting the reliability, unreliability, and mean time to error of a health care professional performing time-continuous tasks, in which he/she can make two types of errors: critical and noncritical. The model state space diagram is shown in Figure 11.2. The numerals in boxes and circle denote the states of the health care professional. The other symbols used in the diagram are defined subsequently.

The model is subject to the following two assumptions:

- Errors occur independently.
- Critical and noncritical error rates are constant.

The following symbols are associated with the model:

i is the ith state of the health care professional; $i = 0$ means the health care professional is performing his/her task normally, $i = 1$ means

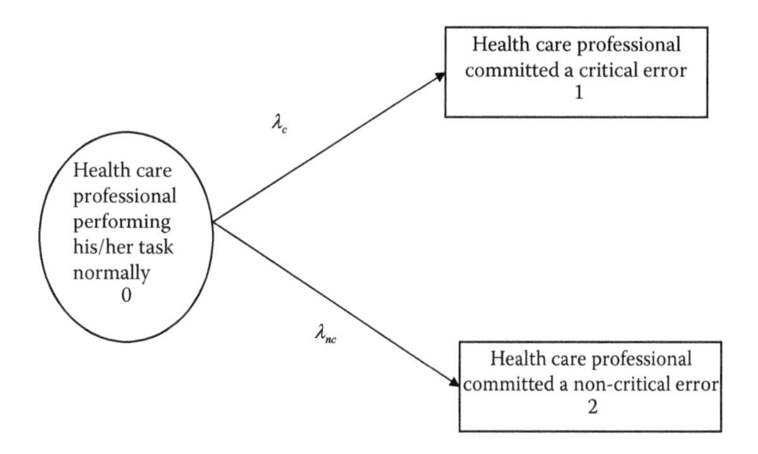

FIGURE 11.2
Model III state space diagram.

the health care professional committed a critical error, $i = 2$ means the health care professional committed a noncritical error.

$P_i(t)$ is the probability of the health care professional being in state i at time t, for $i = 0, 1, 2$.

λ_c is the constant critical error rate of the health care professional.

λ_{nc} is the constant noncritical error rate of the health care professional.

By using the Markov method presented in Chapter 4, we write down the following set of equations for the Figure 11.2 diagram [2,12]:

$$\frac{dP_0(t)}{dt} + (\lambda_{nc} + \lambda_c)P_0(t) = 0 \tag{11.3}$$

$$\frac{dP_1(t)}{dt} - \lambda_c P_0(t) = 0 \tag{11.4}$$

$$\frac{dP_2(t)}{dt} - \lambda_{nc} P_0(t) = 0 \tag{11.5}$$

At time $t = 0$, $P_0(0) = 1$, $P_1(0) = 0$, and $P_2(0) = 0$.

By solving Equations (11.3)–(11.5) using Laplace transforms, we get

$$P_0(t) = e^{-(\lambda_c + \lambda_{nc})t} \tag{11.6}$$

$$P_1(t) = \frac{\lambda_c}{\lambda_c + \lambda_{nc}}[1 - e^{-(\lambda_c + \lambda_{nc})t}] \tag{11.7}$$

$$P_2(t) = \frac{\lambda_{nc}}{\lambda_c + \lambda_{nc}}[1 - e^{-(\lambda_c + \lambda_{nc})t}] \tag{11.8}$$

The reliability and unreliability of the health care professional are given by Equations (11.9) and (11.10), respectively.

$$\begin{aligned} RH_{cp}(t) &= P_0(t) \\ &= e^{-(\lambda_c + \lambda_{nc})t} \end{aligned} \tag{11.9}$$

and

$$URH_{cp}(t) = P_1(t) + P_2(t) = 1 - e^{-(\lambda_c + \lambda_{nc})t} \quad (11.10)$$

where

$RH_{cp}(t)$ = the reliability of the health care professional at time t.

$URH_{cp}(t)$ = the unreliability of the health care professional at time t.

By integrating Equation (11.9) over the time interval $[0,\infty]$, we get the following equation for the health care professional's mean time to error [2,12]:

$$MTTEH_{cp} = \int_0^\infty e^{-(\lambda_c + \lambda_{nc})t} dt = \frac{1}{\lambda_c + \lambda_{nc}} \quad (11.11)$$

where

$MTTEH_{cp}$ = the mean time to error of the health care professional.

Example 11.3

Assume that a health care professional is performing a certain task and his/her critical and noncritical error rates are 0.005 errors per hour and 0.08 errors per hour, respectively. Calculate his/her probabilities of making critical and noncritical errors for a 10-hour mission.

By substituting the given data values into Equation (11.7), we get

$$P_1(10) = \frac{0.005}{(0.005 + 0.08)}[1 - e^{-(0.005+0.08)(10)}] = 0.0337$$

Similarly, by inserting the specified data values into Equation (11.8), we get

$$P_2(10) = \frac{0.08}{(0.005 + 0.08)}[1 - e^{-(0.005+0.08)(10)}] = 0.5389$$

Thus the health care professional's probabilities of making critical and noncritical errors are 0.0337 and 0.5389, respectively.

Example 11.4

A health care professional is performing a certain task and he/she can make two types of errors: critical errors and noncritical errors. Calculate the health care professional's mean time to error, if his/her critical and noncritical error rates are 0.0007 errors per hour and 0.006 errors per hour, respectively.

By substituting the given data values into Equation (11.11), we get

$$MTTEH_{cp} = \frac{1}{(0.0007 + 0.006)}$$

$$= 149.25 \text{ hours}$$

Thus the health care professional's mean time to error is 149.25 hours.

11.6.4 Model IV

This model is concerned with predicting the reliability, unreliability, and mean time to error of a health care professional performing his/her tasks in fluctuating environments (i.e., normal and stressful). Under this scenario, the occurrence of human errors can vary quite significantly—more specifically, from an individual's normal work environment to stressful work environment and vice versa. Thus, this mathematical model is used to estimate the health care professional's reliability, unreliability, and mean time to error under the stated work conditions.

The state space diagram of the model is shown in Figure 11.3. The numerals in circles and boxes denote the health care professional's states.

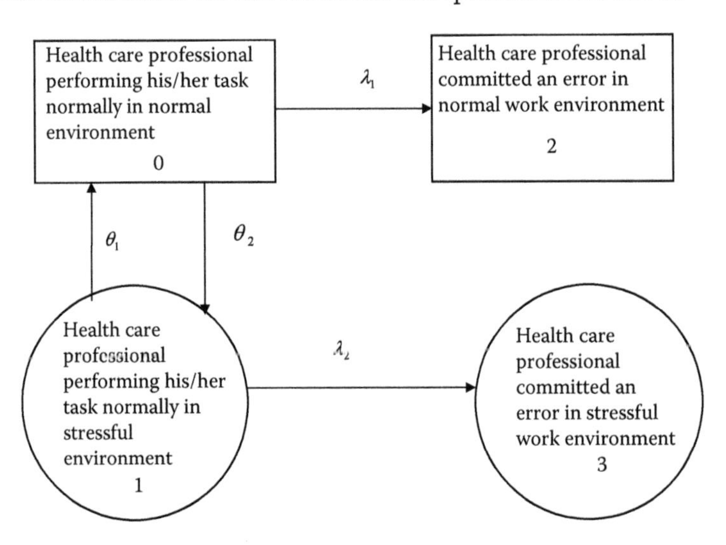

FIGURE 11.3
Model IV state space diagram.

The following three assumptions are associated with the model:

- Error rates of the health care professional are constant.
- The rates of change of the work condition of the health care professional from normal to stressful and vice versa are constant.
- All errors occur independently.

The following symbols are associated with the model:

i is the ith state of the health care professional; $i = 0$ means the health care professional is performing his/her task normally in normal environment, $i = 1$ means the health care professional is performing his/her task normally in stressful environment, $i = 2$ means the health care professional committed an error in normal work environment, $i = 3$ means the health care professional committed an error in stressful work environment.

$P_i(t)$ is the probability of the health care professional being in state i at time t, for $i = 0, 1, 2, 3$.

λ_1 is the constant error rate of the health care professional in normal work environment.

λ_2 is the constant error rate of the health care professional in stressful work environment.

θ_1 is the constant transition rate from stressful work environment to normal work environment.

θ_2 is the constant transition rate from normal work environment to stressful work environment.

By using the Markov method presented in Chapter 4, we write down the following set of equations for the Figure 11.3 diagram [2,12,13]:

$$\frac{dP_0(t)}{dt} + (\lambda_2 + \theta_2)P_0(t) = P_1(t)\theta_1 \tag{11.12}$$

$$\frac{dP_1(t)}{dt} + (\lambda_2 + \theta_1)P_0(t) = P_0(t)\theta_2 \tag{11.13}$$

$$\frac{dP_2(t)}{dt} = P_0(t)\lambda_1 \tag{11.14}$$

$$\frac{dP_3(t)}{dt} = P_1(t)\lambda_2 \tag{11.15}$$

At time $t = 0$, $P_0(0) = 1$, $P_1(0) = 0$, $P_2(0) = 0$, and $P_3(0) = 0$.

By solving Equations (11.12)–(11.15) using Laplace transforms, we get

$$P_0(t) = (x_2 - x_1)^{-1}[(x_2 + \lambda_2 + \theta_1)e^{x_2 t} - (x_1 + \lambda_2 + \theta_1 e^{x_1 t})] \tag{11.16}$$

where

$$x_1 = [-A_1 + \sqrt{A_1^2 - 4A_2}\,]/2 \tag{11.17}$$

$$x_2 = [-A_1 - \sqrt{A_1^2 - 4A_2}\,]/2 \tag{11.18}$$

$$A_1 = \lambda_1 + \lambda_2 + \theta_1 + \theta_2 \tag{11.19}$$

$$A_2 = \lambda_1(\lambda_2 + \theta_1) + \theta_2\lambda_2 \tag{11.20}$$

$$P_2(t) = A_4 + A_5 e^{x_2 t} - A_6 e^{x_1 t} \tag{11.21}$$

where

$$A_3 = \frac{1}{x_2 - x_1} \tag{11.22}$$

$$A_4 = \lambda_1(\lambda_2 + \theta_1) / x_1 x_2 \tag{11.23}$$

$$A_5 = A_3(\lambda_1 + A_4 x_1) \tag{11.24}$$

$$A_6 = A_3(\lambda_1 + A_4 x_2) \tag{11.25}$$

$$P_1(t) = \theta_2 A_3 (e^{x_2 t} - e^{x_1 t}) \tag{11.26}$$

$$P_3(t) = A_7[(1 + A_3)(x_1 e^{x_2 t} - x_2 e^{x_1 t})] \tag{11.27}$$

where

$$A_7 = \lambda_2\theta_2 / x_1 x_2 \tag{11.28}$$

The reliability of the health care professional is given by

$$RH_{cp}(t) = P_0(t) + P_1(t) \tag{11.29}$$

where
$RH_{cp}(t)$ = the reliability of the health care professional at time t.

Similarly, unreliability of the health care professional is expressed by

$$URH_{cp}(t) = P_2(t) + P_3(t) \tag{11.30}$$

where
$URH_{cp}(t)$ = the unreliability of the health care professional at time t.

By integrating Equation (11.29) over the time interval $[0, \infty]$, we get the following equation for the health care professional's mean time to error [2, 12, 13]:

$$\begin{aligned} MTTEH_{cp} &= \int_0^\infty [P_0(t) + P_1(t)]dt \\ &= (\lambda_2 + \theta_2 + \theta_1) / [\lambda_1(\lambda_2 + \theta_1) + \theta_2\lambda_2] \end{aligned} \tag{11.31}$$

where
$MTTEH_{cp}$ = the mean time to error of the health care professional.

Example 11.5

A health care professional is performing his/her tasks under fluctuating normal and stressful environments. The constant transition rates from normal to stressful environment and vice versa are 0.001 times per hour and 0.005 times per hour, respectively. The constant error rates of the professional in normal and stressful environments are 0.007 errors per hour and 0.009 errors per hour, respectively. Calculate his/her mean time to error by using Equation (11.31).

By substituting the given data values into Equation (11.31), we get

$$\begin{aligned} MTTEH_{cp} &= (0.009 + 0.001 + 0.005) / [0.007(0.009 + 0.005) + (0.001)(0.009)] \\ &= 140.187 \text{ hours} \end{aligned}$$

Thus the mean time to error of the health care professional is 140.187 hours.

11.7 Problems

1. Write an essay on the types of patient safety organizations.
2. What are the common functions of patient safety organizations?
3. What are the approaches used by the patient safety organizations to reduce the occurrence of adverse events?
4. What are the governmental patient safety organizations? Describe at least two such organizations.
5. Describe the following three patient safety-related organizations:
   - Australian Patient Safety Foundation
   - National Quality Forum
   - Joint Commission on Accreditation of Healthcare Organizations
6. List at least 10 organizations considered useful to obtain patient safety-related data.
7. Assume that the error rate of a health care professional, in performing his/her assigned tasks, is 0.005 errors per hour. Calculate his/her reliability for an 8-hour mission.
8. Assume that a health care professional's error rate is 0.009 errors per hour. Calculate his/her mean time to error.
9. A health care professional is performing a certain task and his/her critical and noncritical error rates are 0.0006 errors per hour and 0.008 errors per hour, respectively. Calculate the professional's probabilities of making critical and noncritical errors for an 8-hour mission.
10. Assume that a health care professional is performing his/her tasks under fluctuating normal and stressful environments. The constant transition rates from normal to stressful environment and vice versa are 0.002 times per hour and 0.006 times per hour, respectively. The constant error rates of the professional in normal and stressful environments are 0.008 errors per hour and 0.009 errors per hour, respectively. Calculate his/her mean time to error.

11.8 References

1. Kohn, L. T., Corrigan, J. M., Donaldson, M. S., Editors, *To Err Is Human: Building a Safer Health System*, Institute of Medicine Report, National Academy Press, Washington, D.C., 1999.
2. Dhillon, B. S., *Human Reliability: With Human Factors*, Pergamon Press, New York, 1986.

3. Dhillon, B. S., *Reliability Technology, Human Error, and Quality in HealthCare*, CRC Press, Boca Raton, Florida, 2008.
4. *Patient Safety*, World Health Organization, Geneva, Switzerland, 2004. Available online at http://www.who.int/patientsafety/en/, Retrieved July 15, 2006.
5. The National Guideline Clearinghouse, Agency for Healthcare Research and Quality, Rockville, Maryland.
6. Obesity Surgery Complication Rates Higher over Time, Press Release, Agency for Healthcare Research and Quality, Rockville, Maryland, July 24, 2006.
7. National Patient Safety Agency, Press Release, National Health Service, London, UK, 2005.
8. National Quality Forum, 601 13th Street NW, Suite 500 North, Washington, D.C.
9. Wilson, R. M., Van DerWeyden, M. B., The Safety of Australian Healthcare: 10 Years after QAHCS, *Medical Journal of Australia*, Vol. 182, No. 6, 2005, pp. 260–261.
10. Canadian Patient Safety Institute, Suite 1414, 10235-101 Street, Edmonton, Alberta, Canada.
11. The Joint Commission on Accreditation of Healthcare Organizations (JCAHO), 1 Renaissance Blvd., Oakbrook Terrace, Illinois.
12. Dhillon, B. S., *Design Reliability: Fundamentals and Applications*, CRC Press, Boca Raton, Florida, 1999.
13. Dhillon, B. S., Stochastic Models for Predicting Human Reliability, *Microelectronics and Reliability*, Vol. 25, 1985, pp. 729–752.

Appendix: Bibliography—Literature on Patient Safety

A.1 Introduction

Over the years, a large number of publications on patient safety have appeared in the form of journal articles, books, conference proceedings articles, technical reports, and so on. This appendix presents an extensive list of selective publications related, directly or indirectly, to patient safety.

The period covered by the listing is from 1967 to 2011. The main objective of this listing is to provide readers with sources for obtaining additional information on patient safety.

A.2 Publications

Achim, E., Semmer, N. K., Grebner, S., Work Stress and Patient Safety: Observer-Rated Work Stressors as Predictors of Characteristics of Safety-Related Events Reported By Young Nurses, *Ergonomics*, Vol. 49, No. 5–6, 2009, pp. 457–469.

Alfredsdottir, H., Bjornsdottir, K., Nursing and Patient Safety in the Operation Room, *Journal of Advanced Nursing*, Vol. 61, No. 1, 2007, pp. 29–37.

Alvarado, C. J., Carayon, P., Schoofs, H. A., Patient Safety Climate (PSC) in Outpatient Surgery Centers—Part Two, *Proceedings of the Human Factors and Ergonomics Society*, 2005, pp. 1464–1468.

Andersen, H. B., Lipczak, H., Ulriksen, I., Experience of Patient Safety Managers and Local Leaders with Handling Incident Reporting, *Proceedings of the European Safety and Reliability Conference*, 2007, pp. 890–895.

Anderson, J. G., Ramanujam, R., Hensel, D., Anderson, M. M., Sirio, C. A., The Need for Organizational Change in Patient Safety Initiatives, *International Journal of Medical Informatics*, Vol. 75, No. 12, 2006, pp. 809–817.

Angermeier, I., Dunford, B., Boss, A. D., Boss, R. W., The Impact of Participative Management Perceptions on Customer Service, Medical Errors, Burnout, and Turnover Intentions, *Journal of Healthcare Management*, Vol. 15, No. 2, 2009, pp. 127–141.

Anon, R., Cooperation Leads to Safety for Hospitals' Patients and Firemen, *Fire Engineering*, Vol. 110, No. 11, 1657, pp. 1120–1121.

Apgar, B., Patient Attitudes about Physician Mistakes, *American Family Physician*, Vol. 55, No. 6, 1997, pp. 2293–2295.

Armstrong, K., Laschinger, H., Wong, C., Workplace Empowerment and Magnet Hospital Characteristics as Predictors of Patient Safety Climate, *Journal of Nursing Care Quality*, Vol. 44, No. 5, 2008, pp. 133–141.

Armutly, M., Foley, M. L., Surette, J., Belzile, E., McCusker, J., Survey of Nursing Perceptions of Medication Administration Practices, Perceived Sources of Errors and Reporting Behaviours, *Healthcare Quarterly*, Vol. 11, 2008, pp. 58–65.

Aron, D. C., Headrick, L. A., Educating Physicians Prepared to Improve Care and Safety Is No Accident: It Requires a Systematic Approach, *Quality and Safety in Health Care*, Vol. 11, 2002, pp. 168–173.

Aronson, J. K., Medication Errors: What They Are, How They Happen, and How to Avoid Them, *QJM*, Vol. 14, 2009, pp. 623–631.

Aspden, P., *Patient Safety: Achieving a New Standard for Care*, National Academies Press, Washington, D.C., 2004.

Badir, A., The Development of Patient Safety in Turkey: Constraints and Limitations, *Journal of Nursing Care Quality*, Vol. 18, No. 3, 2008, pp. 241–247.

Bagian, J. P., Patient Safety: Lessons Learned, *Pediatric Radiology*, Vol. 36, 2006, pp. 287–290.

Bakken, S., Cook, S., Curtis, L., Desjardins, K., Hyun, S., Jenkins, M., John, R., Klein, W. T., Paguntalan, J., Roberts, W. D., Soupios, M., Promoting Patient Safety through Informatics-Based Nursing Education, *International Journal of Medical Informatics*, Vol. 73, No. 7–8, 2004, pp. 581–589.

Balka, E., Doyle-Walters, M., Lecznarowicz, D., FitzGerald, J. M., Technology, Governance and Patient Safety: Systems Issues in Technology and Patient Safety, *International Journal of Medical Informatics*, Vol. 76, 2007, pp. 35–47.

Barber, N., Rawlins, M., Franklin, B. D., Reducing Prescribing Error: Competence, Control and Culture, *Quality and Safety in Health Care*, Vol. 12, 2003, pp. 129–132.

Battles, J. B., Lilford, R. J., Organizing Patient Safety Research to Identify Risks and Hazards, *Quality and Safety in Health Care*, Vol. 12, 2003, pp. 112–117.

Beasley, L. J., Reliability and Medical Device Manufacturing, *Proceedings of the Annual Reliability and Maintainability Symposium*, 1995, pp. 128–131.

Beauregard, K., Patient Safety, Elephants, Chickens, and Mosquitoes, *Plastic Surgical Nursing*, Vol. 26, No. 3, 2006, pp. 123–127.

Behling, D., Industry Profile: Healthcare—Hazards of the Healthcare Profession, *Occupational Health & Safety*, Vol. 62, No. 2, 1993, pp. 54–57.

Benner, P. E., Malloch, K., Sheets, V., *Nursing Pathways for Patient Safety*, Mosby Elsevier, St. Louis, Missouri, 2010.

Berntsen, K. J., Implementation of Patient Centeredness to Enhance Patient Safety, *Journal of Nursing Quality*, Vol. 21, No. 1, 2006, pp. 15–19.

Billings, C. E., Woods, D. D., Human Error in Perspective: The Patient Safety Movement, *Postgraduate Medicine*, Vol. 109, No. 1, 2001, pp. 23–25.

Bogner, M. S., Human Factors, Human Error and Patient Safety Panel, *Proceedings of the Annual Human Factors Society Conference*, Vol. 2, 1998, pp. 1053–1057.

Bonini, P. A., Zoppei, G., Tomassini, G., Improving Quality and Reducing Errors in Laboratory Medicine: Ethical and Scientific Aspects, *Clinical Chemistry & Laboratory Medicine*, Vol. 46, 2008, pp. 126–127.

Bonis, P. A., Pickens, G. T., Rind, D. M, Foster, D. A, Association of a Clinical Knowledge Support System with Improved Patient Safety, Reduced Complications and Shorter Length of Stay among Medicare Beneficiaries in Acute Care Hospitals in the United States, *International Journal of Medical Informatics*, Vol. 77, No. 11, 2003, pp. 745–753.

Bousvaros, G. A., Don, C., Hopps, J. A., An Electrical Hazard of Selective Angiocardiography, *Canadian Medical Association Journal*, Vol. 87, 1962, pp. 286–288.
Brall, A., Human Reliability Issues in Medical Care—A Customer Viewpoint, *Journal of American Medical Association*, Vol. 28, No. 5, 2006, pp. 46–50.
Braun, S., Usability for Medical Devices, *Medical Physics*, Vol. 34, No. 2, 2007, pp. 658–766.
Brennan, P. F., Safran, C., Patient Safety: Remember Who It's Really For, *International Journal of Medical Informatics*, Vol. 73, No. 7–8, 2004, pp. 547–550.
Brennan, T. A., Leape, L. L., Laird, N. M., Incidence of Adverse Events and Negligence in Hospitalized Patients: Results of the Harvard Medical Practice Study I, *The New England Journal of Medicine*, Vol. 324, 1991, pp. 370–376.
Brennan, T. A., Leape, L. L., Laird, N. M., Incidence of Adverse Events and Negligence in Hospitalized Patients: Results of the Harvard Medical Practice Study II, *The New England Journal of Medicine*, Vol. 324, 1991, pp. 377–384.
Brennan, T. A., Localio, R., Leape, L. L., Laird, N. M., Identification of Adverse Events Occurring during Hospitalization, *Annals of Internal Medicine*, Vol. 112, No. 3, 1990, pp. 221–226.
Brennan, T. A., Jocalio, R. J., Laird, N. M., Reliability and Validity of Judgements Concerning Adverse Events Suffered by Hospitalized Patients, *Medical Care*, Vol. 27, No. 12, 1989, pp. 1148–1158.
Brennan, T. A., Leape, L. L., Laird, N. M., Hebert, L., Localio, A. R., Lawthers, A. G., Newhouse, J. P., Weiler, P. C., Hiatt, H. H., Incidence of Adverse Events and Negligence in Hospitalized Patients, *The New England Journal of Medicine*, Vol.. 324, 1991, pp. 370–376.
Browning, R. A, The Labor Shortage, Patient Safety, and Length of Stay: New Era of Change Agents Prompts Process Improvements through Lab Automation, *JALA*, Vol. 9, No. 1, 2004, pp. 24–27.
Bruner, J. M. R., Hazards of Electrical Apparatus, *Anesthesiology*, Vol. 28, No. 2, 1967, pp. 396–425.
Bruner, J. M. R., Leonard, P. F., *Electricity, Safety, and the Patient*, Year Book Medical Publishers, Chicago, 1989.
Burland, E. M. J., An Evaluation of a Fall Management Program in a Personal Care Home Population, *Healthcare Quarterly*, Vol. 11, No. 3, 2008, pp. 137–140.
Burns, K. K., Canadian Patient Safety Champions: Collaborating on Improving Patient Safety, *Healthcare Quarterly*, Vol. 11, No. 3, 2008, pp. 95–100.
Bushy, A., Roloff, D., A Focus on Patient Safety in an Anesthesiology Department, *Journal of Nursing Quality Assurance*, Vol. 3, No. 1, 1988, pp. 37–45.
Byrd, J., Thompson, L., "It's Safe to Ask": Promoting Patient Safety through Health Literacy, *Healthcare Quarterly*, Vol. 11, No. 3, 2008, pp. 91–94.
Byrns, G., Healthcare Hazard Control & Safety Management, *Professional Safety*, Vol. 51, No. 6, 2006, pp. 46–47
Camishion, R. C., Electrical Hazards in the Research Laboratory, *Journal of Surgical Research*, Vol. 6, 1966, pp. 221–227.
Carayon, P., *Handbook of Human Factors and Ergonomics in Health Care and Patient Safety*, Lawrence Erlbaum Associates, Mahwah, New Jersey, 2007.
Carayon, P., Schoofs, H. A., Alvarado, C. J., Springman, S. R., Ayoub, P., Patient Safety in Outpatient Surgery: The Viewpoint of the Healthcare Providers, *Ergonomics*, Vol. 49, No. 5–6, 2006, pp. 470–485.

Carrie, L., *Understanding Patient Safety*, Quay Books, London, 2007.

Chang, D. C., Handly, N., Abdullah, F., Efron, D. T, Haut, E. R, Haifer, A. H., Pronovost, P. J., Cornwell, E. E., The Occurrence of Potential Patient Safety Events among Trauma Patients, Are They Random?, *Annals of Surgery*, Vol. 247, No. 2, 2008, pp. 327–334.

Chao, C. C., Jen, W. Y., Hung, M., Li, Y. C, Chi, Y. P., An Innovative Mobile Approach for Patient Safety Services: The Case of a Taiwan Healthcare Provider, *Technovation*, Vol. 27, No. 6–7, 2007, pp. 342–351.

Chia, C. C., Wen, Y. J., Yan, P. C., Binshan, L., Improving Patient Safety with RFID and Mobile Technology, *International Journal of Electronic Healthcare*, Vol. 3, No. 2, 2007, pp. 175–192.

Chien, C., Hwang, B., Lin, T., Wang, C., Chong, F., Using Web Technique in the Managing Regulatory Requirements of Medical Equipment for the Nursing Department, *Proceedings of the 28th EMBS Annual International Conference*, 2006, pp. 6773–6776.

Chisholm, L. A., Telder, R., Dolan, A. M., A Patient Safety Program for Small Hospitals, *Biomedical Sciences Instrumentation*, Vol. 10, 1974, pp. 125–128.

Chu, T., Jin, M., Xu, L., Chiang, J., Kao, C., Portable Patient Information Integration System for Patient Safety Improvement, *Proceedings of the ISPA International Workshops*, 2007, pp. 87–95.

Clancy, C. M., Farqubar, M. B., Sharp, B. A. C., Patient Safety in Nursing Practice, *JNCQ*, Vol. 20, No. 3, 2005, pp 193–197.

Clarkson, P. J., Buckle, P., Coleman, R., Stubbs, D., Ward, J., Jarrett, J., Lane, R., Bound, J., Design for Patient Safety: A Review of the Effectiveness of Design in the UK Health Service, *Journal of Engineering Design*, Vol. 15, No. 2, 2004, pp. 123–40.

Classen, D., Patient Safety, The Name is Quality, *Trustee*, Vol. 53, No. 9, 2000, pp. 12–15.

Cone, L. C., Kagan, R. J., Gottschlich, M. M., Enhancing Patient Safety: The Effect of Process Improvement on Bedside Fluoroscopy Time Related to Nasoduodenal Feeding Tube Placement in Pediatric Burn Patients, *Journal of Burn Care & Research*, Vol. 30, No. 4, 2009, pp. 1–6.

Cook, R. I., Woods, D. D., A Tale of Two Stories: Contrasting Views of Patient Safety, *Report from a Workshop on Assembling the Scientific Basis for Progress on Patient Safety*, National Health Care Safety Council of the National Patient Safety Foundation at the AMA, Chicago, Illinois, 2000.

Cortès, P., Krishman, S. M., Lee, I., Goldman, M., Improving the Safety of Patient Controlled Analgesia Infusions with Safety Interlocks and Closed-Loop Control, *Biomedical Instrumentation & Technology*, Vol. 37, No. 8, 2004, pp. 169–175.

Croskerry, P., et al, Editors, *Patient Safety in Emergency Medicine*, Wolters Kluwer Health/Lippincott Williams and Wilkins, Philadelphia, 2009.

Cziraki, K., Lucas, J., Rogers, T., Page, L., Zimmerman, R., Hauer, L. A., Daniels, C., Communication and Relationship Skills for Rapid Response Teams at Hamilton Health Sciences, *Healthcare Quarterly*, Vol. 11, 2008, pp. 66–71.

Dalton, G. D., Samaropoulos, X. F., Dalton, A. C., Improvements in the Safety of Patient Care Can Help End the Medical Malpractice Crisis in the United States, *Health Policy*, Vol. 86, 2008, pp. 153–162.

Damberg, C. L., Ridgely, M. S., Shaw, R., Robin, C. M., Sorbero, M. E. S., Bradley, L. A., Farley, D. O., Adopting Information Technology to Drive Improvements in Patient Safety: Lessons from the Agency for Healthcare Research and Quality Information Technology Grantees, *Health Research and Educational*, Vol. 44. No. 2, 2008, pp. 684–700.

Dana, M. G., Turcsanyi, B., Becich, M. J., Gupta, D., Gilbertson, J. R., Raab, S. S., Database Construction for Improving Patient Safety by Examining Pathology Errors, *American Society for Clinical Pathology (Journal)*, Vol. 124, 2005, pp. 500–509.

Dankelman, J., Grimbergen, C. A., Stassen, H. G., *Engineering for Patient Safety: Issues in Minimally Invasive Procedures*, Lawrence Erlbaum Associates, Mahwah, New Jersey, 2005.

David, C., David, W. B., Health Information Exchange and Patient Safety, *Journal of Biomedical Informatics*, Vol. 40, No. 6, 2007, pp. 40–45.

Davies, B., Special Feature Medical Safety Systems: A Safe Communication System for Wheelchair-Mounted Medical Robots, *Computing & Control Engineering Journal*, Vol. 22, No. 12, 2009, pp. 49–55.

Davies, H. T., Nutley, S. M., Mannion, R., Organizational Culture and Quality of Health Care, *QSHC*, Vol. 9, 2000, pp. 111–119.

Davis, J. M., Strunin, L., Anaesthesia in 1984: How Safe Is It?, *Canadian Medical Association Journal*, Vol. 131, 1984, pp. 437–441.

Davis, J. W., Hoyt, D. B., McArdle, M. S., The Significance of Critical Care Errors in Causing Preventable Death in Trauma Patients in a Trauma Station, *Journal of Trauma*, Vol. 31, 1991, pp. 813–819.

De Lemos, Z., FMEA Software Program for Managing Preventive Maintenance of Medical Equipment, *Biomedical Instrumentation & Technology*, Vol. 37, No. 8, 2004, pp. 162–168.

DeAnda, A., Gaba, M., Unplanned Incidents during Comprehensive Anesthesia Simulation, *Anesthesia and Analgesia*, Vol. 71, 1990, pp. 77–82.

Deboer, G. E., Maurer, W. G., Rhoton, M. F., Education and Patient Safety: Increasing Resident Awareness of Common Anesthetic Misadventures, *Anesthesiology*, Vol. 69, No. 3A, 1988, pp. A796–A797.

Delaney, K. R., Johnson, M. E., Impatient Psychiatric Nursing: Why Safety Must Be Key Deliverable, *Archives of Psychiatric Nursing*, Vol. 22, No. 6, 2008, pp. 386–388.

DerGurahian, J., Not Much Progress, *Modern Healthcare*, Vol. 39, No. 16, 2009, pp. 8–9.

Dieckmann, P., Reddersen, S., Wehner, T., Rall, M., Prospective Memory Failures as an Unexplored Threat to Patient Safety: Results from a Pilot Study Using Patient Simulators to Investigate the Missed Execution of Intentions, *Ergonomics*, Vol. 49, No. 5–6, 2006, pp. 526–543.

Dierks, M. M., Christian, K., Roth, E. M., Sheridan, T. B., Healthcare Safety: The Impact of Disabling "Safety" Protocols, *IEEE Transactions on Systems, Man & Cybernetics*, Vol. 34, No. 6, 2004, pp. 693–698.

Dobbie, A. K., Patient Safety—Class III Equipment Advantages, *Bio-Medical Engineering*, Vol. 8, No. 1, 1973, pp. 125–133.

Donaldson, L., Championing Patient Safety: Going Global, *Quality and Safety in Health Care*, Vol. 11, 2002, pp. 112.

Donaldson, L., Patient Safety: Global Momentum Builds, *Quality and Safety in Health Care*, Vol. 13, 2004, pp. 86.

Dubois, R. W., Brook, R. H., Preventable Deaths: Who, How, and Why?, *Annals of Internal Medicine*, Vol. 109, 1988, pp. 582–589.

Dulworth, S., Case Management and Patient Safety: Opportunities for Process and Performance Improvement, *Archives of Internal Medicine,* Vol. 160, 2005, pp. 69–72.

Duric, N., Littrup, P., Poulo, L., Babkin, A., Pevzner, R., Holsapple, E., Rama, O., Glibe, C., Detection of Breast Cancer with Ultrasound Tomography: First Results with the Computed Ultrasound Risk Evaluation (CURE) Prototype, *Medical Physics,* Vol. 34, No. 2, 2007, pp. 773–785.

Dyer, D., Bouman, B., Davey, M., Ismond, P. K., An Intervention Program to Reduce Falls for Adult In-Patients Following Major Lower Limb Amputation, *Healthcare Quarterly*, Vol. 11, 2008, pp. 117–120.

Eberthard, D. P., Stinebring, R. C., Qualification of High Reliability—Medical Grade Batteries, *Proceedings of the Annual Reliability and Maintainability Symposium,* 1989, pp. 356–362.

Editorial, Promoting Patient Safety by Preventing Medical Error, *Journal of American Medical Association,* Vol. 280, No. 16, 1998, pp. 1144–1148.

Eichhorn, J. H., Prevention of Intraoperative Anesthesia Accidents and Related Severe Injury through Safety Monitoring, *Anesthesiology*, Vol. 70, 1989, pp. 572–577.

Eidesen, K., Aven, T., An Evaluation of Risk Assessment as a Tool to Improve Patient Safety and Prioritize the Resources, *Proceedings of the European Safety and Reliability Conference*, 2007, pp. 171–177.

Eisenberg, J. M., Medical Errors and Patient Safety: A Growing Research Priority, *Health Services Research,* Chicago, Illinois, Vol. 35, No. 3, 2000, pp. XI–XVI.

Elahi, B. J., Safety and Hazard Analysis for Software-Controlled Medical Devices, *Proceedings of the 6th Annual IEEE Symposium on Computer-Based Medical Systems,* 1993, pp. 10–15.

Erlen, J. A., Patient Safety, Error Reduction, and Ethical Practice, *Orthopaedic Nursing,* Vol. 26, No. 2, 2007, pp. 130–133.

Ernst, E. A., MacKrell, T. N., Pearson, J. D., Gutter, G., Wagenknecht, L., Patient Safety: A Comparison of Open and Closed Anesthesia Circuits, *Anesthesiology*, Vol. 67, No. 3A, 1987, pp. A474–A475.

Ervin, N., Does Patient Satisfaction Contribute to Nursing Care Quality?, *The Journal of Nursing Administration*, Vol. 36, No. 3, 2006, pp. 126–130.

Esmail, R., Duchscherer, G., Giesbrecht, J., King, J., Ritchie, P., Zuege, D., Prevention of Ventilator-Associated Pneumonia in the Calgary Health Region: A Canadian Success Story, *Healthcare Quarterly*, Vol. 11, No. 3, 2008, pp. 129–136.

Fabre, J., *Smart Nursing: Nurse Retention and Patient Safety Improvement Strategies*, Springer, New York, 2009.

Fagerhaugh, S. Y., *Hazards in Hospital Care: Ensuring Patient Safety*, Jossey-Bass, San Francisco, 1987.

Farley, D. O., Battles, J. B., Evaluation of the AHRQ Patient Safety Initiative: Framework and Approach, *Health Research and Educational*, Vol. 44, No. 2, 2009, pp. 628–644.

Farley, D. O., Damberg, C. L., Evaluation of the AHRQ Patient Safety Initiative: Synthesis of Findings, *Health Research and Educational*, Vol. 44, No. 2, 2009, pp. 756–776.

Feng, X., Bobay, K., Weiss, M., Patient Safety Culture in Nursing: A Dimensional Concept Analysis, *Journal of Advanced Nursing*, Vol. 63, No. 3, 2007, pp. 310–319.

Ferguson, S. L., To Err Is Human: Strategies for Ensuring Patient Safety and Quality When Caring for Children, *Journal of Pediatric Nursing*, Vol. 16, No. 6, 2001, pp. 438–440

Ferman, J., Medical Errors Spur New Patient-Safety Measures, *Healthcare Executive*, March 2000, pp. 55–56.

Flatley, B. P., Charles, S., Patient Safety: Remember Who It's Really For, *International Journal of Medical Informatics*, Vol. 73, No. 7–8, 2004, pp. 547–550.

Fleming, M., Wentzell, N., Patient Safety Culture Improvement Tool: Development and Guidelines for Use, *Healthcare Quarterly*, Vol. 11, 2008, pp. 10–15.

Frankel, A., Gandhi, T. K., Bates, D. W., Improving Patient Safety across a Large Integrated Health Care Delivery System, *International Journal for Quality in Health Care*, Vol. 15, No. 1, 2003, pp. 131–139.

Freihrr, G., Safety Is Key to Product Quality, Productivity, *Medical Device & Diagnostic Industry Magazine*, Vol. 19, No. 4, 1997, pp. 18–19.

Fries, R. C., Pienkowski, P., Jorgens, J., Safe, Effective and Reliable Software Design and Development for Medical Devices, *Medical Instrumentation*, Vol. 30, No. 2, 1996, pp. 75–80.

Frush, K. S., Alton, M., Frush, D. P., Development and Implementation of a Hospital-Based Patient Safety Program, *Pediatric Radiology*, Vol. 36, 2006, pp. 291–298

Frush, K. S., Fundamentals of a Patient Safety Program, *Pediatric Radiology*, Vol. 38, No. 4, 2008, pp. 685–689.

Fukuda, H., Imanaka, Y., Hayashida, K., Cost of Hospital-Wide Activities to Improve Patient Safety and Infection Control: A Multi-Centre Study in Japan, *Health Policy*, Vol. 87, 2008, pp. 100–111.

Fukuda, H., Imanaka, Y., Hirose, M., Hayashida, K., Factors Associated with System-Level Activities for Patient Safety and Infection Control, *Health Policy*, Vol. 89, 2008, pp. 26–36.

Fulton, J. S., A Clinical Nurse Specialist Agenda for Patient Safety, *Clinical Nurse Specialist*, Vol. 20, No. 2, 2006, pp. 53–54.

Gaba, D. M., Anaesthesiology as a Model for Patient Safety in Healthcare, *British Medical Journal*, Vol. 320, 2000, pp. 785–788.

Gaba, D. M, Human Performance Issues in Anaesthesia Patient Safety, *Problems in Anaesthesia*, Vol. 5, 1991, pp. 329–330.

Galt, K. A., Paschal, K. A., *Foundations in Patient Safety for Health Professionals*, Jones and Bartlett, Sudbury, Massachusetts, 2011.

Gavin-Dreschnack, D., Nelson, A., Fitzgerald, S., Harrow, J., Sanchez-Anguiano, A., Ahmed, S., Powell-Cope, G., Wheelchair-Related Falls: Current Evidence and Directions for Improved Quality Care, *Journal of Nursing Quality*, Vol. 20, No. 2, 2005, pp. 119–127.

Gearge, B., Polly, C., Everlyn, S., Implementation of a Patient Safety Collaborative Forum Facilitates Organizational Learning from Medical Error, *Proceedings of the Human Factors and Ergonomics Society Conference*, 2006, pp. 944–948.

Glance, L. G., Li, Y., Osler, T. M., Mukamel, D. B., Dick, A, W., Impact of Date Stamping on Patient Safety Measurement in Patients Undergoing CABG: Experience with the AHRQ Patient Safety Indicators, *BMC Heath Research*, Vol. 8, 2008, pp. 176–182.

Glenn, B., An Overview of the Patient Safety Movement in Healthcare, Plastic Surgical Nursing, *American Society for Healthcare Risk Management*, Vol. 26, No. 3, 2006, pp. 785–793.

Gorman, C., The Disturbing Case of the Cure That Killed the Patient, *Time,* April, 3, 1995, pp. 60–61.

Gowen, L. D., Specifying and Verifying Safety-Critical Software Systems, *Proceedings of the IEEE 7th Symposium on Computer-Based Medical Systems*, 1994, pp. 235–240.

Gowen, L. D., Yap, M. Y., Traditional Software Development's Effects on Safety, *Proceedings of the 6th Annual IEEE Symposium on Computer-Based Medical Systems*, 1993, pp. 58–63.

Graban, M., *Lean Hospitals: Improving Quality, Patient Safety, and Employee Satisfaction*, CRC Press, Boca Raton, Florida, 2009.

Grant, L. J., Regulations and Safety in Medical Equipment Design, *Anaesthesia*, Vol. 53, 1998, pp. 1–3.

Gravenstein, J. S., How Does Human Error Affect Safety in Anesthesia?, *Surgical Oncology Clinics of North America*, Vol. 9, No. 1, 2000, pp. 81–95.

Greenberg, M. D., Haviland, A. M., Hao, Y., Farley, D. O., Safety Outcomes in the United States: Trends and Challenges in Measurement, *Health Research and Educational*, Vol. 44, No. 2, 2008, pp. 739–755 .

Grinney, R., Patient Safety in Biomedical Engineering, *Electron*, Vol. 49, No. 49, 1974, pp. 33–42.

Gunn, I. P., Patient Safety and Human Error: The Big Picture, *CRNA*, Vol. 11, No. 1, 2000, pp. 41–48.

Habas, P. A., Zurada, J. M., Elmaghraby, A. S., Tourassi, G. D., Reliability Analysis Framework for Computer-Assisted Medical Decision Systems, *Medical Physics*, Vol. 34, No. 2, 2007, pp. 763–772.

Hall, L. M., *Quality Work Environments: For Nurse and Patient Safety*, Jones and Bartlett, Sudbury, Massachusetts, 2005.

Hallock, M. L., Alper, S. J., Karsh, B., A Macro-Ergonomic Work System Analysis of the Diagnostic Testing Process in an Outpatient Health Care Facility for Process Improvement and Patient Safety, *Ergonomics*, Vol. 49, No. 5–6, 2006, pp. 544–566.

Hanada, E., Itoga, S., Takano, K., Kudou, T., Investigations of the Quality of Hospital Electric Power Supply and the Tolerance for Medical Electric Devices to Voltage Dips, *Journal of Medical Systems*, Vol. 31, 2007, pp. 219–223.

Healy, J., Dugdale, P., *Patient Safety First: Responsive Regulation in Healthcare*, Allen and Unwin, Crows Nest, New South Wales, Australia, 2009.

Heyman, B., *Risk, Safety, and Clinical Practice: Healthcare through the Lens of Risk*, Oxford University Press, Oxford, UK, 2010.

Hilborne, L. H., Setting the Stage for the Second Decade of the Era of Patient Safety: Contributions by the Agency for Healthcare Research and Quality and Grantees, *Health Research and Educational*, Vol. 44, No. 2, 2009, pp. 623–627.

Holland, R., Patient Safety, Complications and Consequences, *Current Opinion in Anaesthesiology*, Vol. 3, 1990, pp. 864–868.

Hopps, J. A., A Rationale for Patient Electrical Safety, *Proceedings of the Biological Engineering Society 15th International Conference on Recent Advances in Biomedical Engineering*, 1975, pp. 17–22.

Hopps, J. A., Electrical Hazards in Hospital Instrumentation, *Proceedings of the Annual Symposium on Reliability*, 1969, pp. 303–307.

Howanitz, P. J., Errors in Laboratory Medicine—Practical Lessons to Improve Patient Safety, *Archives of Pathology & Laboratory Medicine*, Vol. 129, 2005, pp. 1252–1261.

Howard, F., Technology and Pediatric Patient Safety: What to Target Is the Dilemma, *Journal of Pediatrics*, Vol. 152, 2008, pp. 153–155.

Hu, Y., Feng, Y. M., Wang, M. S., Lu, W. W., Chip-Microcomputer Based Safety Tester for Medical Equipment, *Proceedings of the 20th Annual International Conference of the IEEE Engineering in Medicine and Biology Society*, 1998, pp. 3364–3366.

Hubbell, R. C., Thatcher, R. K., Increasing Patient Safety through an Automated Blood Processing System, *Proceedings of the 4th Annual Symposium on Computer Applications in Medical Care*, 1980, pp. 1261–1263.

Huey, M. T., Height of Hospital Beds and Inpatient Falls—A Threat to Patient Safety, *JONA*, Vol. 37, No. 12, 2000, pp 537–538.

Hughes, G. R., Clancy, M. C., Working Conditions That Support Patient Safety, *Journal of Nursing Care Quality*, Vol. 20, No. 4, 2005, pp. 289–292.

Hurwitz, B., Sheikh, A., Editors, *Health Care Errors and Patient Safety*, Wiley-Blackwell/BMJ, Hoboken, New Jersey, 2009.

Hyndman, B., H., A Thirty-Two Year Perspective on a Clinical Engineer's Contribution to Patient Safety, *Proceedings of the Annual International Conference of the IEEE Engineering in Medicine and Biology*, 2007, pp. 710–716.

IEC 601-1: *Safety of Medical Electrical Equipment, Part l: General Requirements*, International Electrotechnical Commission (IEC), Geneva, 1977.

Irurita V. F., Factors Affecting the Quality of Nursing Care: The Patient's Perspective, *International Journal of Nursing Practice*, Vol. 5, 1999, pp. 86–94.

Jackson, L. D., Little, J., Kung, E., Williams, E. M., Siemiatkowska, K., Plowman, S., Safe Medication Swallowing in Dysphagia: A Collaborative Improvement Project, *Healthcare Quarterly*, Vol. 11, No. 3, 2008, pp. 110–116.

James, B., Margaret, A., Technology and Patient Safety: A Two-Edged Sword, *Biomedical Instrumentation and Technology*, Vol. 36, No. 2, 2002, pp. 84–88.

Jenicek, M., *Medical Error and Harm: Understanding, Prevention, and Control*, Productivity Press, New York, 2011.

Johnson, B. W., Aylor, J. H., Reliability and Safety Analysis in Medical Applications of Computer Technology, *Proceedings of the Annual Reliability and Maintainability Symposium*, 1988, pp. 96–100.

Johnson, V. R., Hummel, J., Kinninger, T., Lewis, R. F, Immediate Steps toward Patient Safety, *Healthcare Financial Management*, Vol. 58, No. 2, 2004, pp. 56–61.

Jones, D. S., Do You Know Quality When You See It?, *Journal of Health Care Compliance*, Vol. 32, No. 7, 2009, pp. 43–73.

Kaelber, D. C., Bates, D. W., Health Information Exchange and Patient Safety, *Journal of Biomedical Informatics*, Vol. 40, 2007, pp. 40–45.

Kaye, R., Crowley, J., *Medical Device Use-Safety: Incorporating Human Factors Engineering into Risk Management*, CDRH, Office of Health and Industry Programs, U.S. Department of Health and Human Services, Washington, D.C., 2000.

Kelley, J., Medical Diagnostic Device Reliability Improvement and Prediction Tools—Lessons Learned, *Proceedings of the Annual Reliability and Maintainability Symposium*, 1999, pp. 29–31.

Keselman, A., Patel, V. L., Johnson, T. R., Zhang, J., Institutional Decision-Making to Select Patient Care Devices: Identifying Venues to Promote Patient Safety, *Journal of Biomedical Informatics*, Vol. 36, No. 1–2, 2003, pp. 31–44.

Khalafalla, A. S., Electrical Safety for Medical Equipment and Suggested Patient Shielding for Monitoring or Diagnostic Purposes, *Proceedings of the 26th Annual Conference on Engineering in Medicine and Biology*, 1973, pp. 429.

Khatri, N., Baveja, A., Boren, S. A., Mammo, A., Medical Errors and Quality of Care: From Control to Commitment, *California Management Review*, Vol. 48, No. 3, 2006, pp. 116–135.

Kim, J., Kyungeh, A., Kim, M. K., Yoon, S. H., Nurse's Perception of Error Reporting and Patient Safety Culture in Korea, *Western Journal of Nursing Research*, Vol. 29, 2007, pp. 827–833.

Klinger, A. R., Negative Resistance Patient Isolators—Some Considerations, *Journal of Clinical Engineering*, Vol. 2, No. 4, 1977, pp. 332–335.

Klomp, A. M., Lucas, J. H. M., Considerations on the Safety of Electrical Equipment in Medical Practice for the Preparation of an International Standard, *Eurocon*, Vol. 2, No. 71, 1971, pp. 18–22.

Kostopoulou, O., From Cognition to the System: Developing a Multilevel Taxonomy of Patient Safety in General Practice, *Ergonomics*, Vol. 49, No. 5–6, 2006, pp. 486–502.

Krause, T. R., Hidley, J. H., *Taking the Lead in Patient Safety: How Healthcare Leaders Influence Behavior and Create Culture*, John Wiley and Sons, Hoboken, New Jersey, 2009.

Krenzischek, D. A., Clifford, T. L., Windle, P. E., Mamaril, M., Patient Safety: PeriAnesthesia Nursing's Essential Role in Safe Practice, *Journal of PeriAnesthesia Nursing*, Vol. 22, No. 6, 2007, pp. 385–392.

Landro, L., For Patients, a List of Hospital Hazards, *The Wall Street Journal* (Eastern edition), December 25, 2008, p. D.2

Landro, L., The Informed Patient: Hospitals Reuse Medical Devices to Lower Costs, *The Wall Street Journal*, March 19, 2008, p. D1.

Landro, L., Wireless Medical Items Pose Risks, *Wall Street Journal* (Eastern edition), June 25, 2008, p. B7.

Latino, R. J., *Patient Safety: The PROACT Root Cause Analysis Approach*, CRC Press, Boca Raton, Florida, 2009.

Laura, L., Human Error in Patient Controlled Analgesia: Incident Reports and Experimental Evaluation, *Proceedings of the Human Factors and Ergonomics Society Conference*, Vol. 2, 1998, pp. 1043–1047.

Le Cocq, A. D., Application of Human Factors Engineering in Medical Product Design, *Journal of Clinical Engineering*, Vol. 12, No. 4,1987, pp. 271–277.

Leape, L. L, Woods, D. D., Hatlie, M. J., Kizre, K. W., Promoting Patient Safety by Preventing Medical Error, *Journal of American Medical Association*, Vol. 280, No. 16, 1998, pp. 1444–1448.

Leape, L. L., Swankin, D. S., Yessian, M. R., A Conversation on Medical Injury, *Public Health Reports*, Vol. 114, 1999, pp. 302–317.

Leape, L. L., The Preventability of Medical Injury, in *Human Error in Medicine*, M. S. Bogner, Ed., Lawrence Erlbaum Associates, Hillsdale, New Jersey, 1994, pp. 13–27.

Leitgeb, N., *Safety in Electromedical Technology*, Interpharm Press, Buffalo Grove, Illinois, 1996.

Levkoff, B., Increasing Safety in Medical Device Software, *Medical Device & Diagnostic Industry Magazine*, Vol. 18, No. 9, 1996, pp. 92–97.

Lewis, R. Q., Fletcher, M., Implementing a National Strategy for Patient Safety: Lessons from the National Health Service in England, *Quality and Safety in Health Care*, Vol. 14, 2005, pp. 135–139.

Li, P., Schneider, J. E., Ward, M. M., Effect of Critical Access Hospital Conservation on Patient Safety, *Health Research and Educational*, Vol. 42, No. 6, 2007, pp. 2089–2108.

Lifvergren, S., Chakunashvili, A., Bergman, B., Docherty, P., Online Statistical Monitoring of Critical Patient Data Increases Patient Safety, *Proceedings of the Portland International Center for Management of Engineering and Technology Conference*, 2008, pp. 895–899.

Lin, L., Liang, B. A., Addressing the Nursing Work Environment to Promote Patient Safety, *Nursing Forum*, Vol. 42, No. 1, 2007, pp. 20–30.

Lin, L., Vicente, K. J., Doyle, D. J., Patient Safety, Potential Adverse Drug Events, and Medical Device Design: A Human Factors Engineering Approach, *Journal of Biomedical Informatics*, Vol. 34, No. 4, 2001, pp. 274–284.

Lipton, M. J., Allen, R., Ream, A. K., Hyndman, B. H., A Conductive Catheter to Improve Patient Safety during Cardiac Catheterization, *Circulation*, Vol. 58, No. 6, 1978, pp. 1190–1195.

Livenson, A. R., Electrical Equipment in Medical Institutions and Problems of Safety, *Biomedical Engineering*, Vol. 13, No. 1, 1979, pp. 4–8.

Loivisto, E., Safety Aspects in the Standardization of Electromedical Equipment, *Proceedings of the Annual International Conference on Biomedical Transducers*, 1975, pp. 457–462.

Ludbrook, G. L., Webb, R. K., Fox, M. A., Singleton, R. J., Problems before Induction of Anaesthesia: An Analysis of 2000 Incident Reports, *Anaesthesia and Intensive Care*, Vol. 21, 1993, pp. 593–595.

Ludbrook, G. L., Webb, R. K., Fox, M. A., The Australian Incident Monitoring Study: Physical Injuries and Environmental Safety in Anaesthesia: An Analysis of 2000 Incident Reports, *Anaesthesia and Intensive Care*, Vol. 21, No. 5, 1993, pp. 659–663.

Lunn, J., Devlin, H., Lessons from the Confidential Inquiry into Preoperative Deaths in Three NHS Regions, *Lancet*, Vol. 2, 1987, pp. 1384–1386.

Macintosh, M. A., Choo, C. W., Information Behavior in the Context of Improving Patient Safety, *Journal of the American Society for Information Science and Technology*, Vol. 56, No. 12, 2005, pp. 1332–1345.

Marconi, M., Sirchia, G., Increasing Transfusion Safety by Reducing Human Error, *Current Opinion in Haematology*, Vol. 7, No. 6, 2000, pp. 382–386.

Marcus, M. L., Biersach, B. R., Regulatory Requirements for Medical Equipment, *Proceedings of the Annual Reliability and Maintainability Symposium*, 2003, pp. 23–29.

Martin, J. L., Norris, B. J., Murphy, E., Crowe, J. A., Medical Device Development: The Challenge for Ergonomics, *Applied Ergonomics*, Vol. 39, 2007, pp. 271–283.

Matayoshi, M., Saito, M., Electrical Safety and Reliability of Health Care Facilities Equipped with Class I Equipment—Safety Limit for Occurrence of Single-Fault Conditions and Its Maintenance, *Japanese Journal of Medical Electronics and Biological Engineering*, Vol. 18, No. 2, 1980, pp. 105–111.

Matlow, A., Stevens, P., Urmson, L., Wray, R., Improving Patient Safety through a Multi-Faceted Internal Surveillance Program, *Healthcare Quarterly*, Vol. 11, No. 3, 2008, pp. 101–109.

Mattie, A., Ben-Chitrit, R., Patient Safety Legislation: A Look at Health Policy Development, *Policy, Politics, & Nursing Practice*, Vol. 8, No. 4,2007, pp. 251–261.

McFadden, K. L., Stock, G. N., Gowen, C. R., Implementation of Patient Safety Initiatives in US Hospitals, *International Journal of Operations & Production Management*, Vol. 26, No. 3, 2006, pp. 326–347.

McKeon, L., M., Cunningham, P., D., Oswaks, D., Jill, S, Improving Patient Safety: Patient-Focused, High-Reliability Team Training, *Journal of Nursing Quality*, Vol. 25, 2005, pp. 87–92.

McLinn, J. A., Reliability Development and Improvement of a Medical Instrument, *Proceedings of the Annual Reliability and Maintainability Symposium*, 1996, pp. 236–242.

Meiris, D. C., Clarke, J. L., Nash, D. B., Culture Change at the Source: A Medical School Tackles Patient Safety, *American Journal of Medical Quality*, Vol. 21, 2006, pp. 9–12.

Mendel, P., Damberg, C. L., Sorbero, M., E. S., Varda, D. M., Farley, D. O., The Growth of Partnerships to Support Patient Safety Practice Adoption, *Health Research and Educational*, Vol. 44, No. 2, 2008, pp. 717–737.

Merali, R., Orser, B. A., Leeksma, A., Lingard, S., Belo, S., Hyland, S., Medication Safety in the Operating Room: Teaming Up to Improve Patient Safety, *Healthcare Quarterly*, Vol. 11, 2008, pp. 54–57.

Meyer, G., Battles, J., Hart, J. C., Tang, N., The US Agency for Healthcare Research and Quality's Activities in Patient Safety, *International Journal for Quality in Health Care*, Vol. 15, No. 1, 2003, pp. 125–130.

Miksta, J. A., Industry Response to Patient Safety Concerns, *Journal of Enterostomal Therapy*, Vol. 13, No. 5, 1986, pp. 180–181.

Millenson, M. L., Pushing the Profession: How the News Media Turned Patient Safety into a Priority, *Quality and Safety in Health Care*, Vol. 11, 2002, pp. 57–63.

Miller, M. R., Elixhauser, A., Zhan C., Meyer, G. S., Patient Safety Indicators: Using Administrative Data to Identify Potential Patient Safety Concerns, *Health Service Research*, Vol. 36, No. 5, 2001, pp. 110–132.

Miller, M. R., Elixhauser, A., Zhan, C., Patient Safety Events during Pediatric Hospitalizations, *Pediatrics*, Vol. 111, 2003, pp. 1358–1366.

Miller, M. R., Zhan, C., Pediatric Patient Safety in Hospitals: A National Picture in 2000, *Pediatrics*, Vol. 113, 2004, pp. 1741–1746.

Minter, S. G., Is Healthcare Safety Being Neglected?, *Occupational Hazards*, Vol. 61, No. 4, 1999, pp. 146–151.

Mitchell, P., H., Patient-Centered Care—A New Focus on a Time-Honored Concept, *The Journal of the American Academy of Nursing*, Vol. 12, No. 34, 2008, pp. 45–52.

Mojdehbakhsh, R., Tsai, W. T., Kirani, S., Elliott, L., Retrofitting Software Safety in an Implantable Medical Device, *IEEE Software*, No. 1, Jan. 1994, pp. 41–50.

Momtahan, K., Burns, C. M., Jeon, J., Hyland, S., Gabriele, S., Using Human Factors Methods to Evaluate the Labeling of Injectable Drugs, *Healthcare Quarterly*, Vol. 11, No. 3, 2008, pp. 122–128.

Morell, R. C., Eichhorn, J. H., *Patient Safety in Anaesthetic Practice*, Churchill Livingston, New York, 1997.

Muntlin, A., Gunningberg, L., Carlsson, M., Patients' Perceptions of Quality of Care at an Emergency Department and Identification of Areas for Quality Improvement, *Journal of Clinical Nursing*, Vol. 15, 2006, pp. 1045–1056.

Murff, H. J., Patel, V. L., Hripcsak, G., Bates, D. W., Detecting Adverse Events for Patient Events for Patient Safety Research: A Review of Current Methodologies, *Journal of Biomedical Informatics*, Vol. 36, No. 1–2, 2003, pp. 131–43.

Myles, S. C., Monitoring Patient Safety at the Bedside, *Journal of Nursing Quality Assurance*, Vol. 3, No. 1, 1988, pp. 57–62.

Nakajima, K., Kurata, Y., Takeda, H., A Web-Based Incident Reporting System and Multidisciplinary Collaborative Projects for Patient Safety in a Japanese Hospital, *Quality and Safety in Health Care*, Vol. 14, 2005, pp. 123–129.
Nakhleh, R. E., Patient Safety and Error Reduction in Surgical Pathology, *Archives of Pathology & Laboratory Medicine*, Vol. 132, 2008, pp. 181–185.
Nash, D. B., Goldfarb, N. I., *The Quality Solution: The Stakeholder's Guide to Improving Health Care*, Jones and Bartlett, Sudbury, Massachusetts, 2006.
Nelson, A., *Safe Patient Handling and Movement: A Guide for Nurses and Other Care Providers*, Springer, New York, 2006.
Neudorf, K., Dyck, N., Scott, D., Dick, D. D., Nursing Education: A Catalyst for the Patient Safety Movement, *Healthcare Quarterly*, Vol. 11, 2008, pp. 35–39.
Nevland, J. G., Electrical Shock and Reliability Considerations in Clinical Instruments, *Proceedings of the Annual Symposium on Reliability*, 1969, pp. 308–313.
Newhouse, R. P., Poe, S., *Measuring Patient Safety*, Jones and Bartlett, Sudbury, Massachusetts, 2005.
Ng, W. S., Tan, C. K., On Safety Enhancement for Medical Robots, *Reliability Engineering and System Safety*, Vol. 54, 1996, pp. 35–45.
Nickerson, T., Jenkins, M., Greenall, J., Using ISMP Canada's Framework for Failure Mode and Effects Analysis: A Tale of Two FMEAs, *Healthcare Quarterly*, Vol. 11, 2008, pp. 40–46.
Nicklin, W., Mass, H., Affonso, D. D., O'Connor, P., Ferguson-Paré, M., Jeffs, L., Tregunno, D., White, P., Patient Safety Culture and Leadership within Canada's Academic Health Science Centres: Towards the Developments of a Collaborative Position Paper, *Nursing Leadership*, Vol. 17, No. 1, 2004, pp. 22–34.
Nigam, R., Mackinnon, N. J., David, U., Hartnell, N. R., Levy, R. A., Gurnham, E. M., Nguyen, T. T., Development of Canadian Safety Indicators for Medication Use, *Healthcare Quarterly*, Vol. 11, 2008, pp. 47–53.
Nolan, T. W., System Changes to Improve Patient Safety, *British Medical Journal*, Vol. 320, 2000, pp. 771–773.
Odom-Forren, J., Patient Safety: Nursing Priority, *Journal of PeriAnesthesia Nursing*, Vol. 22, No. 6, 2007, pp. 446–448.
Olivier, D. P., Software Safety: Historical Problems and Proposed Solutions, *Medical Device and Diagnostic Industry Magazine*, Vol. 17, No. 7, 1995, pp. 116–124.
Ortiz, E., Meyer, G., Burstin, H., The Role of Clinical Informatics in the Agency for Healthcare Research and Quality's Efforts to Improve Patient Safety, *JAMA*, Vol. 280, 2001, pp. 508–512.
Ortiz-Posadas, M. R., Vernet-Saavedra, E. A., Electrical Safety Priority Index for Medical Equipment, *Proceedings of the 28th IEEE EMBS Annual International Conference*, 2006, pp. 6614–6616.
Otsuka, Y., Ayzawa, J., Akiyoshi, M, Noguchi, H., Improvement Framework for Safety Rule Using Incident Reporting and Worker's Practical Heuristics (Empirical Consideration in Patient Safety Management), *Transactions of the Japan Society of Mechanical Engineers*, Vol. 74, No. 4, 2008, pp. 1012–1019.
Panescu, D., Emerging Technologies: Design and Medical Safety of Neuromuscular Incapacitation Devices, *Proceedings of the 20th Annual International Conference of the IEEE Engineering in Medicine and Biology*, 2009, pp. 57–67.
Paulsen, P. K., Nissen, T., Patient Safety Unit for a Hot-Film Anemometer, Used for Blood-Velocity Determination in Humans, *Medical & Biology Engineering & Computing*, Vol. 20, 1982, pp. 625–627.

Pawlson, G., O'Kane, M. E., Malpractice Prevention, Patient Safety, and Quality of Care: A Critical Linkage, *The American Journal of Managed Care*, Vol. 10, No. 4, 2004, pp. 281–284.

Pedreira, M. L., G, Marin, H. F., Patient Safety Initiatives in Brazil: A Nursing Perspective, *International Journal of Medical Informatics*, Vol. 73, No. 7–8, 2004, pp. 563–567.

Pei-Chung, L., Liu, L., Frank, K., Ming-Hui, J., Developing a Patient Safety Based RFID Information System: An Empirical Study in Taiwan, *Proceedings of the Annual International Conference on Management of Innovation and Technology*, 2006, pp. 55–61.

Pelczarski, K. M., Heyman, J., Laying the Foundation for Preventing Device-Related Patient Safety Problems, *Biomedical Instrumentation & Technology*, Vol. 42, No. 3, 2008, pp. 225–228.

Pernice, A., How to Increase the Safety of the Patient, *Technology Elettriche*, Vol. 7, No. 8, 1980, pp. 50–59.

Pesanka, D. A., Greenhouse, P. K., Rack, L. L., Delucia, G. A., Perret, R. W., Scholle, C. C., Johnson, M. S., Ticket to Ride: Reducing Handoff Risk during Hospital Patient Transport, *Journal of Nursing Care Quality*, Vol. 18, No. 3, 2008, pp. 217–223.

Peters, G. A., Peters, B. J., *Medical Error and Patient Safety: Human Factors in Medicine*, CRC Press, Boca Raton, Florida, 2008.

Peterson, M. G. E., Privacy versus Safety: Who Is Safe?, *Proceedings of the 15th IEEE Symposium on Computer-Based Medical Systems*, 2002, pp. 346–352.

Poe, S. S., Patient Safety Planting the Seed, *Journal of Nursing Quality*, Vol. 20, No. 3, 2005, pp. 198–202.

Porter, M., Gerrish, P., Tyler, L., Murray, S., Mauriello, R., Soto, F., Phetteplace, G., Hareland, S., Reliability Considerations for Implantable Medical ICs, *Proceedings of the 46th Annual International Reliability Physics Symposium*, 2008, pp. 516–552.

Preboth, M., Medication Errors in Paediatric Patients, *American Family Physician*, Vol. 63, No. 2, 2001, pp. 678.

Quigley P. A., Hahm, B., Collazo, S., Gibson, W., Janzen, S., Powell-Cope, G., Rice, F., Sarduy, I., Tyndall, K., White, S. V., Reducing Serious Injury from Falls in Two Veterans' Hospital Medical-Surgical Units, *Journal of Nursing Care Quality*, Vol. 24, No. 1, 2009, pp. 33–41.

Raab, S. S., Improving Patient Safety through Quality Assurance, *Archives of Pathology & Laboratory Medicine*, Vol. 130, 2006, pp. 633–637.

Raab, S. S., Marie, D., Zarbo, R. J., Meier, F. A., Geyer, S. J., Jensen, C., Anatomic Pathology Databases and Patient Safety, *Archives of Pathology & Laboratory Medicine*, Vol. 129, 2005, pp. 1246–1251.

Rantucci, M. J., Stewart, C., Stewart, I., *Focus on Safer Mediation Practices*, Wolters Kluwer/Lippincott Williams and Wilkins Health, Philadelphia, 2009.

Rapala, K., Nova, J. C., Clinical Patient Safety—Achieving High Reliability in a Complex System, *Digital Human Modeling*, Vol. 54, 2007, pp. 710–716.

Reames, K., Ensuring Patient Safety, *Journal of Nursing Quality Assurance*, Vol. 3, No. 1, 1988, pp. 72–75.

Reducing Errors, Improving Safety, Letters, *British Medical Journal*, Vol. 321, 2000, pp. 505–509.

Reynard, J., Reynolds, J., Stevenson, P., *Practical Patient Safety*, Oxford University Press, New York, 2009.

Ridgway, M., Guidelines for Clinical Engineering Programs, *Journal of Clinical Engineering*, Vol. 6, No. 1, 1981, pp. 53–63.

Rieders, C., Pennsylvania's Patient Safety Authority, *Journal of American College of Radiology*, Vol. 2, No. 8, 2005, pp. 690–695.

Riviera, A. J., Karsh, B., Human Factors and Systems Engineering Approach to Patient Safety for Radiotherapy, *International Journal of Radiation Oncology Biology Physics*, Vol. 71, No. 1, 2008, pp. 174–177.

Robert, D. W., Robert, B. H., Preventable Deaths: Who, How Often, and Why? *Annals of Internal Medicine*, Vol. 109, 1998, pp. 582–589.

Robson, R., Pelletier, E., Disclosure, Apology and Early Compensation Discussions after Harm in the Healthcare Setting, *Healthcare Quarterly*, Vol. 11, No. 3, 2008, pp. 85–90.

Rose M., Drake, D., Baker, G., Watkins, F., Waters, W., Pokorny, M., Caring for Morbidly Obese Patients: Safety Considerations for Nurse Administrators, *Dimensions of Critical Care Nursing*, Vol. 23, No. 2, 2008, pp. 76–80.

Ruland, C. M., Improving Patient Safety through Informatics Tools for Shared Decision Making and Risk Communication, *International Journal of Medical Informatics*, Vol. 73, No. 7, 2004, pp. 551–557.

Runciman, W. B., Sellen, A., Webb, R. K., Williamson, J. A., Current, M., Errors, Incidents and Accidents in Anaesthetic Practice, *Anaesthesia and Intensive Care*, Vol. 21, No. 5, 1993, pp. 506–519.

Russell, C., Human Error: Avoidable Mistakes Kill 100,000 Patients a Year, *Washington Post*, February 18, 1992.

Rust, T. B., Wagner, M. L, Hoffman, C., Rowe, M., Neumann, I., Broadening the Patient Safety Agenda to Include Safety in Long-Term Care, *Healthcare Quarterly*, Vol. 11, 2008, pp. 31–34.

Saltos, R., Man Killed by Accident with Medical Radiation, *Boston Globe*, June 20, 1986, p. 1.

Sandars, J., Cook, G., *ABC of Patient Safety*, Blackwell/BMJ, Malden, Massachusetts, 2007.

Santamour, B., Aiming for Perfection, *Quality Care*, Vol. 22, No, 34, 2009, pp. 13–16.

Santel, C., Trautmann, C., Liu, W., The Integration of a Formal Safety Analysis into the Software Engineering Process: An Example from the Pacemaker Industry, *Proceedings of the Symposium on the Engineering of Computer-Based Medical Systems*, 1988, pp. 152–154.

Saufl, N. M., 2007 National Patient Safety Goals, *Journal of PeriAnesthesia Nursing*, Vol. 22, No. 2, 2007, pp. 125–127.

Savage, G. T., Ford, E. W., *Patient Safety and Healthcare Management*, Emerald JAI, Bingley, UK, 2008.

Sayre, K., Kenner, J., Jones, P., Safety Models: An Analytical Tool for Risk Analysis of Medical Device Systems, *Journal of Biomedical Engineering*, Vol. 4, 1982, pp. 445–451.

Scheffler, A. L., Zipperer, L., Patient Safety, *Proceeding of the Symposium on the Enhancing Patient Safety and Reducing Errors in Health care*, Rancho Mirage, California, 1998, pp. 115–120.

Schlich, T., Trohler, U., *The Risks of Medical Innovation: Risk Perception and Assessment in Historical Context*, Routledge, New York, 2006.

Scholefield, H., Embedding Quality Improvement and Patient Safety at Liverpool; Women's NHS Foundation Trust, *Best Practice & Research Clinical Obstetrics and Gynecology*, Vol. 21, No. 4, 2007, pp. 593–607.

Schuster, P. M., Nykolyn, L., *Communication for Nurses: How to Prevent Harmful Events and Promote Patient Safety*, F. A. Davis, Philadelphia, 2010.

Schutz, A. L., Counte, M., Assessment of Patient Safety Research from an Organizational Ergonomics and Structural Perspective, *Ergonomics*, Vol. 50, No. 9, 2007, pp. 1451–1484.

Scott A., California Hospital Embraces Smart Pump Technology, *Drug Topics*, Vol. 11, No. 152, 2008, pp. 18–19.

Sedman, A., Harris, J. M., Schulz, K., Schwalenstocker, E., Remus, D., Scanlon, M., Bahl, V., Relevance of the Agency for Healthcare Research and Quality Patient Safety Indicators for Children's Hospitals, *Pediatrics*, Vol. 115, 2005, pp. 135–145.

Sharpe, V. A., *Accountability: Patient Safety and Policy Reform*, Georgetown University Press, Washington, D.C., 2004.

Shepherd, M., *A Systems Approach to Medical Device Safety*, Monograph, Association for the Advancement of Medical Instrumentation, Arlington, Virginia, 1983.

Shinn, J. A., Root Cause Analysis: A Method of Addressing Errors and Patient Risk, *Progress in Cardiovascular Nursing*, Vol. 15, No. 1, 2000, pp. 25–25.

Singer, S., Shoutzu, L., Falwell, A., Gaba, D., Baker, L., Relationship of Safety Climate and Safety Performance in Hospitals, *Health Research and Educational*, Vol. 44, No. 2, 2008, pp. 399–421.

Sloane, E. B., Gehlot, V., Ensuring Patient Safety by Using Colored Petri Net Simulation in the Design of Heterogeneous, Multi-Vendor, Integrated, Life-Critical Wireless (802.x) Patient Care Device Networks, *Proceedings of the 27th Annual International Conference of the IEEE Engineering in Medicine and Biology*, 2006, pp. 98–108.

Slonim, A. D., Marcin, J. P, Turenne, W., Hall, M., Joseph, J. G., Pediatric Patient Safety Events during Hospitalization: Approaches to Accounting for Institution-Level Effects, *Health Research and Educational*, Vol. 42, No. 6, 2007, pp. 2275–2293.

Smith, C. E., Peel, D., Safety Aspects of the Use of Microprocessors in Medical Equipment, *Measurement and Control*, Vol. 21, No. 9, 1988, pp. 275–276.

Smith, R. B., Robyn, C., Pamela O., Mark, W., Simpson, L., Medicaid Markets and Pediatric Patient Safety in Hospitals, *Health Research and Educational*, Vol. 42, No. 5, 2007, pp. 1981–1998.

Sorbero, M. E. S., Ricci, K. A., Lovejoy, S., Haviland, A. M., Smith, L., Bradley, L. A., Hiatt, L., Farley, D. O., Assessment of Contributions to Patient Safety Knowledge by the Agency for Healthcare Research and Quality-Funded Patient Safety Projects, *Health Research and Educational*, Vol. 44, No. 2, 2008, pp. 646–664.

Spath, P., *Engaging Patients as Safety Partners: A Guide for Reducing Errors and Improving Satisfaction*, Health Forum, Chicago, 2008.

Spath, P., *Error Reduction in Health Care: A Systems Approach to Improving Patient Safety*, AHA Press, Chicago, 2000.

Stanley, P. E., Monitors That Save Lives Can Also Kill, *Modern Hospital*, Vol. 108, No. 3, 1967, pp. 119–121.

Stetson, J., Patient Safety: Prevention and Prompt Recognition of Regurgitation and Aspiration, *Anesthesia and Analgesia*, Vol. 53, No. 1, 1973, pp. 142–147.

Stroetmann, V. N., Spichtinger, D., Stroetmann, K. A., Thierry, J. P., ICT for Patient Safety: Towards a European Research Roadmap, *Lecture Notes in Bioinformatics*, Springer, Inc., Berlin, 2006, pp. 482–493.

Studer, P., Inderbitzin, D., Surgery-Related Risk Factors, *Current Opinion in Critical Care*, Vol. 22, 2009, pp. 1233–1241.

Sullivan, J. M., Martin, R. H., *Patient Safety Handbook*, ABA, Chicago, 2008.

Taktak, A. F. G, Brown, M. C., Evidence-Based Analysis of Field Testing of Medical Electrical Equipment, *Proceedings of the 28th IEEE EMBS Annual International Conference*, 2006, pp. 4078–4080.

Tardif, G., Aimone, E., Boettcher, C., Fancott, C., Andreoli, A., Velji, K., Implementation of a Safety Framework in a Rehabilitation Hospital, *Healthcare Quarterly*, Vol. 11, 2008, pp. 21–25.

Tavakoli, H., Karami, M., Rezai, J., Esfandiari, K., Khashayar, P., When Renewing Medical Equipment Is Necessary: A Case Report, *International Journal of Health Care Quality Assurance*, Vol. 20, No. 7, 2007, pp. 616–619.

Taylor, E. F., The Effect of Medical Test Instrument Reliability on Patient Risks, *Proceedings of the Annual Symposium on Reliability*, 1969, pp. 328–330.

Taylor, S. L., Ridgely, M. S., Greenberg, M. D., Sorbero, M. E. S., Teleki, S. S., Damberg, C. L., Farley, D. O., Experiences of Agency for Healthcare Research and Quality-Funded Projects That Implemented Practices for Safer Patient Care, *Health Research and Educational*, Vol. 44, No. 2, 2009, pp. 665–683.

Teleki, S. S., Damberg, C. L., Sorbero, M. E. S., Shaw, R. N., Bradley, L. A., Quigley, D. D., Fremont, A. M., Farley, D. O., Training a Patient Safety Work Force: The Patient Safety Improvement Corps, *Health Research and Educational*, Vol. 44, No. 2, 2008, pp. 701–715 .

Thiagarajan, R. R., Bird, G. L., Harrington, K., Charpie, J. R., Ohye, R. C., Steven, J. M., Epstein, M., Laussen, P. C, Improving Safety for Children with Cardiac Disease, *Cardiology in the Young*, Vol. 17, No. 2, 2007, pp. 127–132.

Thompson, D. A., Cowan, J., Hlzmueller, C., Wu, A. W., Bass, E., Pronovost, P., Planning and Implementing a Systems-Based Patient Safety Curriculum in Medical Education, *American Journal of Medical Quality*, Vol. 23, 2008, pp. 271.

Thompson, P. W., Safer Design of Anaesthesia Equipment, *British Journal of Anaesthesia*, Vol. 59, 1987, pp. 913–921.

Thompson, R. C., Fault Therapy Machines Cause Radiation Overdoses, *FDA Consumer*, Vol. 21, No. 10, 1987, pp. 37–38.

Troug, R., *Talking with Patients and Families about Medical Error: A Guide for Education and Practice*, Johns Hopkins University Press, Baltimore, Maryland, 2011.

Tuttle, D., Holloway, R., Baird, T., Sheehan, B., Skelton, W. K., Electronic Reporting to Improve Patient Safety, *Quality and Safety in Health Care*, Vol. 13, 2004, pp. 281–286.

Tzeng, H., Yin, C., Are Call Light Use and Response Time Correlated with Inpatient Falls and Inpatient Dissatisfaction?, *Journal of Nursing Care Quality*, Vol. 24, No. 3, 2009, pp. 232–242.

Tzeng, H., Yin, C., Heights of Occupied Patient Beds: A Possible Risk Factor for Inpatient Falls, *Journal of Clinical Nursing*, Vol. 17, 2008, pp. 1503–1509.

Tzeng, H., Yin, C., No Safety, No Quality: Synthesis of Research on Hospital and Patient Safety (1996–2007), *Journal of Nursing Care Quality*, Vol. 22, No. 4, 2007, pp. 299–306.

Tzeng, H. M., Yin, C. Y., *Patient Unit Safety and Care Quality: Promotion of Self-Healing Systems during Hospital Stays*, Nova Medical Books, New York, 2008.

Van De Castle, B., Kim, J., Pedreira, M. L. G., Paiva, A., Goossen, W., Bates, D. W., Information Technology and Patient Safety in Nursing Practice: An International Perspective, *International Journal of Medical Informatics*, Vol. 73, No. 7–8, 2004, pp. 607–614.

Vance, J. A., *A Guide to Patient Safety in the Medical Practice*, American Medical Association, Chicago, 2008.

Varshney, U., Pervasive Healthcare and Wireless Health Monitoring, *Mobile Network Applications*, Vol. 12, 2007, pp. 113–127.

Velji, K., Baker, R. G., Francott, C., Andreoli, A., Boaro, N., Tardif, G., Aimone, E., Sinclair, L., Effectiveness of an Adapted SBAR Communication Tool for a Rehabilitation Setting, *Healthcare Quarterly*, Vol. 11, 2008, pp. 72–79.

Vincent, C., *Patient Safety*, Elsevier Churchill Livingstone, New York, 2006.

Vincent, C., *Patient Safety: The Heart of Healthcare Quality*, Wiley-Blackwell, Chichester, United Kingdom, 2010.

Wachter, R. M., *Understanding Patient Safety*, McGraw Hill Book Company, New York, 2008.

Wagar, E. A., Tamashiro, L., Yasin, B., Hilborne L., Bruchner D. A., Patient Safety in the Clinical Laboratory: A Longitudinal Analysis of Specimen Identification Errors, *Archives of Pathology & Laboratory Medicine*, Vol. 130, 2006, pp. 1662–1668.

Walsh, K., Antony, J., Improving on Patient Safety and Quality: What Are the Challenges and Gaps in Introducing an Integrated Electronic Adverse Incident and Record System within Health Care Industry?, *International Journal of Health Care Quality Assurance*, Vol. 20, No. 2, 2007, pp. 107–115.

Walshe, K., Boaden, R. J., Editors, *Patient Safety: Research into Practice*, Open University Press, Maidenhead, UK, 2006.

Warholak, T. L., Nau, D. P., *Quality and Safety in Pharmacy Practice*, McGraw Hill Book Company, New York, 2010.

Wear, J. O., Hospital Patient Safety, *Proceedings of the 26th Annual Conference on Engineering in Medicine and Biology*, 1973, pp. 352–366.

Weeks, W. B., West, A. N., Rosen, A. K., Bagian, J. P., Comparing Measures of Patient Safety for Inpatient Care Provided to Veterans within and outside the VA System in New York, *Quality in Health Care*, Vol. 17, 2008, pp. 58–64.

Weinger, M. B., Slagle, J., Jain, S., Ordonez, N., Retrospective Data Collection and Analytical Techniques for Patient Safety Studies, *Journal of Biomedical Informatics*, Vol. 36, No. 1–2, 2003, pp. 106–119.

Weller, C., Radio or Wired Telemetry for Clinical Applications? Patient Safety Standards Improved by Wired Telemetry, *Biotelemetry*, Vol. 2, 1974, pp. 208–210.

Whelpton, D., Equipment Management: The Cinderella of Bio-engineering, *Journal of Biomedical Engineering*, Vol. 10, 1988, pp. 499–505.

Whelpton, D., Roberts, R., Safety of Medical Electrical Equipment and BS5724, *Journal of Biomedical Engineering*, Vol. 4, 1982, pp. 185–196.

White, D., Suter, E., Parboosingh, J., Taylor, E., Communities of Practice: Creating Opportunities to Enhance Quality of Care and Safe Practices, *Healthcare Quarterly*, Vol. 11, No. 3, 2008, pp. 80–84.

Wiig, S., Linde, P. H., Patient Safety in the Interface between Hospital and Risk Regulator, *Proceedings of the European Safety and Reliability Conference*, 2007, pp. 152–158.

Windle, P., E., Reflection on Safety in Perianesthesia Settings, *Journal of PeriAnesthesia Nursing*, Vol. 22, No. 6, 2007, pp. 365–366.

Witters, D., Portnoy, S., Casamento, J., Ruggera, P., Bassen, H., Medical Device EMI: FDA Analysis of Incident Reports, and Recent Concerns for Security Systems and Wireless Medical Telemetry, *Proceedings of the 28th IEEE EMBS Annual International Conference*, 2006, pp. 1289–1291.

Wolff, A., Taylor, S., *Enhancing Patient Care: A Practical Guide to Improving Quality and Safety in Hospitals*, MJA Books, Sydney, Australia, 2009.

Wolosin, R. J., Vercler, L., Matthews, J. L., Am I Safe Here? Improving Patients' Perceptions of Safety in Hospitals, *Journal of Nursing Care Quality*, Vol. 21, No. 1, 2005, pp. 30–38.

Woods, D. D., Cook, R. I., Sarter, N., McDonald, J. S., Mental Models of Anesthesia Equipment Operation: Implications for Patient Safety, *Anesthesiology*, Vol. 71, No 3, 1989, pp. 982–983.

Yang, C., Shaping Up China's Medical Device Industry, *The China Business Review*, Vol. 24, 2008, pp. 24–28.

Yarkey, P., *Medical Quality Management: Theory and Practice*, Jones and Bartlett, Sudbury, Massachusetts, 2010.

Young, V. L., Botney, R., *Patient Safety in Plastic Surgery*, Quality Medical Publishers, St. Louis, Missouri, 2009.

Youngberg, B. J., *Principles of Risk Management and Patient Safety*, Jones and Bartlett, Sudbury, Massachusetts, 2011.

Zane, M., Patient Care Appraisal, *Proceedings of the Annual Reliability and Maintainability Symposium*, 1976, pp. 84–91.

Zhan, C., Miller, M. R., Administrative Data Based Patient Safety Research: A Critical Review, *Quality in Health Care*, Vol. 12, 2003, pp. 158–163.

Zhang, J., Johnson, T. R, Patel, V. L, Paige, D. L, Kubose, T., Using Usability Heuristics to Evaluate Patient Safety of Medical Devices, *Journal of Biomedical Informatics*, Vol. 36, No. 1, 2003, pp. 23–30.

Zimmerman, R., Ivan, I., Christoffersen, E., Shaver, J., Developing a Patient Safety Plan, *Healthcare Quarterly*, Vol. 11, 2008, pp. 26–30.

Zimmerman, R., Ivan, I., Daniels, C., Smith, T., Shaver, J., An Evaluation of Patient Safety Leadership Walkarounds, *Healthcare Quarterly*, Vol. 11, 2008, pp. 16–20.

Index

N

Q

R

S

For Product Safety Concerns and Information please contact our EU representative GPSR@taylorandfrancis.com
Taylor & Francis Verlag GmbH, Kaufingerstraße 24, 80331 München, Germany

www.ingramcontent.com/pod-product-compliance
Ingram Content Group UK Ltd.
Pitfield, Milton Keynes, MK11 3LW, UK
UKHW021613240425
457818UK00018B/540

* 9 7 8 1 1 3 8 1 1 5 0 0 2 *